# Birds, Bats, Bugs, Beavers, Bacteria: Lessons from Nature

## Steven Jakobi

"There lives the dearest freshness,
Deep down things."

-Gerard Hopkins

# Table of contents

# FOREWORD

In early October, 2016, the Committee awarded the year's Nobel Prize in Chemistry to Jean-Pierre Sauvage, Sir J. Fraser Stoddart, and Bernard Feringa. These scientists "...have opened this entire field of molecular machinery," according to Nobel Prize Committee member, Olof Ramstrom. "They have really mastered motion control at the molecular scale," he noted.

National Public Radio correspondent, Camila Domonoske, reported that the "...groundbreaking work on molecular machines... [involved] tiny, man-made structures with moving parts capable of performing tasks... Feringa designed the first molecular motor by making a motor blade to spin in one direction."

These discoveries are, of course, commendable and may lead to lots of new products in the future. But we should also remember that these technologies had already been "invented" by living organisms billions of years ago. Just think of the small, hair-like structures, called cilia, or the larger locomotory appendages, the whip-like flagella, of bacteria, protozoa and other cell types, including the sperm cell. These structures also use "molecular motors that spin in one direction." And the unicellular organism "decides"

1

when and how to activate them, what direction the cell should take in response to various environmental cues, and for how long the action should prevail. On top of that the cell makes its own energy to power such structures.

I am not trying to take anything away from the learned scientists and I am happy for the recognition they so richly deserve. My only point is the same that I made in the introduction to the first edition: eons ago, nature has worked out many of the processes that we are just beginning to understand. And this should give us pause to appreciate nature and the organisms that occupy this planet rather than spending untold amounts of energy to subjugate it to our will. We should do this, if for no other reason, for our own species' preservation. For, if we continue to fight the natural world, nature will ultimately win the contest.

*****

The year, 2017, will mark the 50th anniversary of my immigration to the United States. As a child growing up behind the "Iron Curtain," I could only dream that one day I would live in America. Despite the Communist propaganda (or, perhaps, because of it), I read everything I could get my hands on that dealt with the human character, political systems, and natural history of the North American continent. I read authors like Dreiser, Steinbeck, Irwin Shaw, and Hemingway, but my favorite books were the Leatherstocking series of James Fenimore Cooper. My imagination soared with majestic old-growth forests, exotic birds and mammals, "noble" and "savage" Indians, and heroic soldiers. How could I guess in 1962, at age 11, that I would eventually live near the very areas that Cooper wrote about. I consider myself very fortunate to be able to live and work in this beautiful part of North America whose hills and rivers and people and fauna and flora offer endless fascination and adventure.

****

An earlier version of this book was published in 2015, as *Giorgio the 'Possum and Other Stories from Nature.* The few people who read it were generally complimentary, but also pointed out several errors in that book. I have made an attempt to correct those mistakes or factual errors.

In addition to changing, updating, and expanding several of the previous essays, this book contains eight new stories. I also added 14 pen and ink, pencil, and watercolor illustrations.

As always, each story is short and as nontechnical as I could make it and still retain the essence of the organism. The essays can be read in any order in a few minutes' time. I would love to hear from readers and I hope you will contact me via my website, listed on the inside front cover of the book.

# INTRODUCTION

A few years ago, the director of my college's honors program asked me to address the students at their monthly meeting. "What would you like me to talk about?" I asked. "Whatever you want," he said. This obviously was not very helpful. For the next several days I chewed on the end of my pencil as I tried to put some ideas on the legal pad in front of me.

As the days wore on my quandary deepened and I ended up stealing an idea. I "borrowed" the concept of my talk after listening to one of my favorite radio shows, National Public Radio's *Science Friday*. I don't remember now who the prominent guests were that afternoon, but the topic was "Why engineers should study biology." That was it, then! I knew that most of the students in the honors program were not science majors and, in particular, none of our engineering students ever take my biology classes.

I compiled a list of examples that I thought would be pertinent in such a talk, including the triple-woven designs of the cell wall of plants, and the incredible toughness and flexibility of spider webs as marvels of material strength; the structure of DNA as information storage and retrieval systems; the construction of a woodpecker's skull to keep the brain from getting damaged while drilling holes into wood – a design now being investigated by bicycle and

4

football helmet manufacturers; and the like. I kept the examples simple and my explanations as short and succinct as I could muster and we had a lively questions and answer session to conclude the hour.

The take-home message was this: There is a lot to learn from nature. Many of the technical innovations we continuously "invent" were, in fact, resolved by living systems a long time ago. The helicopter and airplane are poor imitations of what some insects and birds are capable of accomplishing in flight; submarines pale in comparison to the diving abilities of some seafaring creatures; and the brain is still a better tool than any computer. Chemists are still trying to figure out how to imitate the steps of photosynthesis, a process that a single algal cell can perform without any problem. And what man-made machine can do what a heart can: perform its task error-free over 100,000 times per day for many decades? I don't wish to sound like a Luddite and disparage technology. Rather, I nurture the idea that the natural world around us and even its most "primitive" and humble creature can teach us much.

Over the years, I had been invited back several times to talk with members of the Honors Club. We covered a range of topics, from evolution to viral diseases. I also had a chance to visit with garden clubs, forestry groups, members of historical societies, and book clubs where I talked about organic gardening, natural ways of pest control and, frequently, the topic was my research work with chestnut blight, a devastating disease that nearly killed all of our beloved American chestnut trees. At some point it occurred to me that maybe I could reach a wider audience if I took some of the topics of my talks and compiled them in a book.

When I started working on this collection of essays I had no intention of including myself in them. As I began to write, though, personal memories began to intrude into the stories. While not wanting this book to be a literary "selfie," I did recall the first piece of advice given to a would-be writer: "Write what you know about."

I grew up in a big city in the heart of Europe. My family were city folk and an occasional picnic in a park was about as close as they wanted to get to nature. Not me. I was drawn to the outdoors as far back as I can remember. One of my earliest memories is that of sitting in a bed of red and white snapdragons, marveling at a fat green caterpillar emblazoned with a huge yellow and blue eyespot. I remember poking it with sharp stiff pine needles and watching it twist and turn in response to the irritation. At age 2 or 3, I knew nothing about insect metamorphosis, or protective coloration, or the damage that these caterpillars may have caused by their herbivory. I just knew that this was better than any toy I could play with. Even today, after 60 years, the smell of snapdragons brings back a flood of childhood memories and the image of that green caterpillar.

When I was about six years old, my mother moved us to an apartment building that had a small, overgrown back yard with a couple of large Lombardy poplar trees. I couldn't wait for the school bell to ring so I could race home to my "jungle." One of my favorite plants was a woody vine growing along the fence that separated our building from a small house and an auto repair shop. I liked the plant's stout, pleached stem and its small, purple, star-shaped flowers with their creamy yellow centers. On some of the flowers, when mature, the purple parts bent backward to form what to me looked like a crown or a Turk's turban. As the summer progressed the flowers turned into clusters of small red oval berries. Decades later I learned that my plant with the blue-purple turban of petals and yellow stamens was the woody nightshade, *Solanum dulcamara*, a relative of the tomato and potato.

At age 10, I had a chance to spend a summer on a farm. They had no electricity or indoor toilet, but they had pigs, chickens, a cow, and a huge garden. I thought I was in heaven with the rich smell of freshly cut hay, the earthy smell of the garden, the pungent scent of the cow and pig. I learned how to grind corn and to milk the cow, and I was allowed to collect the eggs from the henhouse. My life-long interests in gardening and farming were born during that summer.

In all of these adventures, it didn't matter to me that I often didn't know the names of trees, or why chickens laid eggs every day, or how caterpillars became moths. All I knew was that I was fascinated by all of them and loved them all.

Now, when I tell people what I teach, I get pained grimaces and comments like, "I was never good at biology." Mostly, I think, it is a matter of how we, the instructors, approach the subject that is in large part responsible for these kinds of responses. We are so driven to teach "facts" and rely so much on memorization of meaningless jargon that we cause our audiences' eyes to glaze over and lose interest. To be sure, a certain amount of knowledge of concepts and terminology are necessary in any field, but do we have to take the fun and sense of wonderment right out of the student?

I struggle with this question, just like thousands of my colleagues do. I am constantly trying to strike the right balance, especially for the non-biology majors who take some of my classes.

I hope that the stories I have chosen will be interesting, meaningful and useful for the reader. I hope that, in some small way, I can spark or, perhaps, re-kindle interest in biology. I purposely kept the stories very short and nontechnical. I tried to include history, lore and personal recollections and anecdotes to illustrate that biology touches our everyday lives every single day and in every way.

# KARTOFFELKÄFER!

*Insects...are not curiosities; they are creatures
common with ourselves bound by the laws
of the physical universe...*
*-Robert Evans Snodgrass*

I was born and reared in Budapest, Hungary. The city was, in the 1950s and '60s, a drab, unexciting place behind the Iron Curtain. Bombed out factories, sunken bridges, and empty lots where houses once stood, were still commonplace following the maelstrom and destruction of World War II and then the Revolution of 1956. I was an only child and a loner, and spent many hours after school roaming the hills ringing the Buda side of the city to get away from the diesel fumes of trucks and buses and the clanging and clattering of streetcars. Summer vacations were cherished because I could get away from the city.

There were years when my mother and I would take the train to Lake Balaton, where her state-owned company had summer cottages for its workers near the tiny town of Balatonlelle. Some of my most vivid memories of childhood are from those days of splashing around in the shallow, warm waters of the lake. The gentle breeze, wide open blue sky and sandy beaches are still with me today. And memories of... potato beetles.

Yep. Potato beetles. During most summers, there were tens of thousands of these insects floundering in the water. Sometimes, I rescued some of them and took them to shore. Other times, I made paper boats for them, only to let them drown when the boat sprang a leak and sank. But, mostly, like most 10-year-old boys would, I just cruelly crushed them.

I often wondered: Where had they come from? Why were they floating in the gentle ripples of the lake, helplessly flailing and clearly out of their element? As a small boy, I didn't know the answers but I was fascinated and puzzled at the same time. They left such a deep impression on me that even half a century later I can see them in my mind's eye, smell their peculiar odor, and feel their hard shell between my fingertips.

I asked a teacher back in Budapest about these beetles. He said something like, "Oh, they are the tools of Western imperialism." I didn't know what he meant, but I knew enough not to pursue my inquiry further. I quite forgot about that comment until a few days ago when I stumbled across images of posters on the internet. But, more about that later.

The Colorado potato beetle, *Leptinotarsa decemlineata*, probably originated in northern Mexico or in the southwestern U.S. It was first captured and named in the 1820s in Mexico, where it was found munching on weeds belonging to the Nightshade Family of plants. In 1859, there was a severe outbreak of beetle infestation in cultivated potato fields in Nebraska. The beetle apparently acquired a taste for potatoes and other economically important cultivated Nightshades, including tomato, eggplant, pepper and tobacco.

The insect quickly spread eastward and wherever it showed up it caused crop failure and economic hardship, especially for potato farmers. By the 1880s, the bug was ubiquitous across the U.S. The damage was so severe in many locations that control measures had to be found quickly. In a fashion typical of the unbridled arrogance and confidence of Manifest Destiny after the Civil War, New York newspaper editor Horace Greeley is quoted as saying: "Man is bigger than the potato bug and will master it." "Not so fast," said the beetle.

Europe soon experienced the insect and drastic measures were immediately taken to eradicate it. Infested fields were burned and plowed under and all manner of chemicals were tried. The control measures worked until World War I. However, an infestation near an American military base in France quickly spread to epidemic proportions across Europe. The damage was especially severe in Holland, Belgium and Germany, but no mainland European country was spared. England managed to stave off disaster by strict quarantines and Draconian measures (burning or plowing under fields of potatoes, etc.) wherever the beetle tried to gain a foothold.

The Colorado potato beetle is a small, handsome insect. Its yellow body is lined with ten black-brown stripes, giving it the other common names of "10-lined spearman" or "10-

lined potato beetle." Its species name, *decemlineata*, also literally translates to "ten lines." It is a precocious insect with a voracious appetite. Both larvae and adults eat leaves and quickly defoliate their host plant. Mature beetles are strong fliers and can spread quickly from field to field.

Just before World War II an amazing chemical was (re)discovered and introduced to kill the potato bug and other insects. The pesticide was DDT and initially it worked miracles. Some scientists confidently declared that the scourge of insects would be a thing of the past within ten to 20 years. "Not so fast," said the bugs. By the 1960s many arthropods, including the potato beetle, developed resistance not only to DDT but many other insecticides, as well.

East Germany experienced some of the worst damage from the beetles in the 1950s; nearly half of the potato fields were completely wiped out. Already reeling from inefficient production on "agricultural collectives," the Communist governments of Eastern Block countries would never acknowledge their ineptitude. Instead, their propaganda machines quickly sprang into action with stories and posters of "American imperialism," and "revangist" and "hegemonistic" Western Powers colluding to overthrow these "workers' paradises." The German word for the potato beetle is "Kartoffelkäfer." If the beetles were a tool of American efforts to undermine the East German state – after all, it is the COLORADO potato beetle, right?! – then these dastardly schemes had to be stopped. The beetles were thus depicted in posters draped in the colors of the American flag: armies of "Amikäfers" were met by roadblocks manned by shiny-faced heroic Communist youth at the East German border. Other posters were equally inane. One showed the beetles' "faces" under a magnifying glass with clearly recognizable menacing features of the leaders of Britain, France and West Germany.

As a child, I knew nothing about population biology and how overcrowding and lack of food might send these insects on a desperate search for more food – even if it meant trying to fly across a large lake and perishing by the thousands. I also didn't know about politics and propaganda and the "Cold War." But even at a tender young age, the teacher's remarks about the beetles being tools of Western imperialism made no sense to me. All I knew was that the bugs were in the lake year after year – some years more than others – and that it was "fun" to crunch them. Little did I know that smashing their bodies inadvertently contributed to fighting imperialism and helped to build the great "socialist" Peoples' Republic of Hungary!

# MEMORIES OF WEASELS

# AND BEASTS

*A vast difference clearly exists between the*
*real world ...and the world as perceived*
*by individuals.*
*-Peter Farb*

I had a decision to make after graduating from high school. I could work at the Ford transmission plant in Sharonville or the GM body plant in Norwood, Ohio. Or I could go to college. But if college, then what course of studies to pursue? My heart said, study art. Most everyone tried to talk me out of that. I relented.

In high school I had a wonderful chemistry teacher named Mr. Caskey and I loved his class. So I chose chemistry at the university. I quickly learned that I hated it, especially calculus and analytical chemistry, which in course schedules was always appropriately abbreviated as "anal. chem." I floundered and eventually walked away from my classes. Two years of minimum wage jobs followed. I pumped gas. I bussed tables. I was a security guard. I worked for the city highway department, sweeping streets in the summer and salting roads in the winter.

Upon returning to college I chose biology. I struggled there, as well, as genetics and developmental biology didn't really interest me (much later that would change). I was more interested in the natural sciences and ecology and I did fairly well in those courses.

Glancing at the bulletin board one day in the new science complex, a small, hand-written note piqued my interest:

*"Grad student needs help with field work. No pay but guaranteed rosy cheeks. Please see Rick in room 411."*

I went to see him. It was the beginning of a many-years-long friendship, and an association that changed my attitudes about a whole lot of things.

The work consisted of helping Rick with his doctoral research project. He was studying the ecology and natural history of the long-tailed weasel, *Mustela frenata*. He used a capture-tag-release system to study the weasel population. His field plots were in Boone County, Kentucky, not far from the Greater Cincinnati Airport. We sampled the abundance and distribution of prey (mostly field mice and voles) and attempted to determine the home range and hunting patterns of their predator, the weasel.

In order to study weasels' movements Rick used radio telemetry tracking equipment. He was a wizard with electronics and designed and etched his own circuit boards and built his own radio collars from scratch. To outfit a small, slender but ferocious animal with a well-fitting transmitter was no easy task. First, the animal had to be anesthetized (I don't remember what we used for that) without killing it. Then its neck had to be fitted with a tiny transmitter in such a way that it would still allow the weasel to carry out its normal daily routine when released back into the wild.

Weasels are notoriously clever at slithering into underground burrows through the smallest openings. Thus, it was of paramount importance that the collars had to be securely fastened (we did lose a few over the course of a year). Each transmitter had its own unique signal so that we could tell which animal(s) we were trying to follow at a given time.

The tracking equipment consisted of bulky, hand-held antennas about 2 feet tall and three feet wide, attached to a shoebox-sized receiver. At that time there were no computers, no satellites, no GPS (global positioning system) or GIS (geographic information system) to aid in tracking the animals. Weasels are generally nocturnal hunters and their travels cover fields, edges of forests, and densely wooded areas. Rick and I would try our best to follow for hours the weasels' signals in cold weather or on hot summer evenings, in pitch dark or on moonlit nights. Just two uncoordinated guys stumbling through rutted cornfields and densely wooded areas, getting our bulky antennas snagged on low-lying branches. Not a pretty sight, I'm sure.

Our work routine began in the afternoon by checking traps to sample and estimate the density of the prey population. At dusk, we would turn on the tracking devices and try to find, follow, and map the routes that the weasel(s) traveled. We would crash through the woods and lumber across fields until about midnight, then take a break for a few hours. After a light meal, we would sleep for about 2-3 hours before resuming tracking until early morning.

On a humid, sweltering moonless night, I was sleeping on the ground next to the work truck when I suddenly awakened with the uneasy feeling of one being stalked. I was lying on top of my sleeping bag, covered with a film of dew, and scared out of my mind. I could hear the heavy breathing of a beast just feet away from my head.

Darkness scares humans because we feel so vulnerable, unable to rely on one of our most important senses. At 3 o'clock in the morning all of our worries and fears are magnified, as we are at the lowest ebb of our physiologic functioning. Images of the Hound of Baskerville or a blood-thirsty vampire flashed through my mind (Stephen King's "Cujo" hadn't been published yet at the time). I didn't dare to move (or, perhaps, I couldn't, paralyzed with fear) to get my flashlight. I weakly called to Rick who was happily snoring in the cargo bed of the truck.

Now, I could feel the hot, malodorous breath on my face as the beast circled me. I could sense that it was something large. Were there still wolves in Kentucky? Maybe, a cougar? A rabid raccoon?I was petrified. I was sure that at any moment fangs were going to rip into my throat. Mercifully, Rick finally heard my pitiful croak and switched on his lamp. It was a dog. The neighbor farmer's big old shaggy dog came over to check us out. I sat there shivering in cold sweat on a hot night, and Rick was laughing his head off. I made him swear that he wouldn't tell others about this. Now, forty-some years later, I am spilling the beans.

We tracked weasels for nearly a year but I never got to see Rick's finished work. I moved from Ohio in the autumn of 1976, and he didn't complete his dissertation for another five or six years. But, for me, it was a wonderful time to work outdoors on an enjoyable project, and an experience that allowed me to solidify my decision that biology was what I wanted to pursue for my livelihood. I have been fortunate to work in many biology-related disciplines, including cancer research, in hospital labs as a histotechnologist, as chestnut blight fighter, and as college professor during a four-decades-long career.

# REBIRTH

*Peace with the earth is the first peace.*
*- Henry Beston*

In the early 1970s I lived in my stepfather's house in Ohio. One day a real estate agent came to our house. Whether it was a "cold call" or he had an appointment I do not know but he was hawking parcels in a yet undeveloped part of St. Lucie County, Florida. The building lots for sale were located in what then was still a swamp. As a young and idealistic college student who has just taken a course in ecology, I argued vehemently with the salesman about destruction of the wetland. My self-righteous and sonorous diatribe about displacing native plant and animal species did not go over well. The agent was angry, and my stepfather just rolled his eyes. In the end, my stepfather did not invest in the property. It wasn't because of my impassioned arguments; rather, he thought that this may be a real estate scam.

The salesman wasn't lying. The swamps were drained and homes were built. Port St. Lucie blossomed from a sleepy little hamlet of cattle ranches, citrus groves and swamps with 250 homes in 1961, to a bustling city of 160,000 inhabitants on Florida's Treasure Coast in 2010. The

2010 census recorded 400,000 residents in the surrounding metro area.

The story of Port St. Lucie is not unique. When European settlers arrived in North America in the 1600s, there were an estimated 220 million acres of wetlands and swamps in what now constitutes the United States. By the 1980s, more than half of that area was drained, filled, dammed, or otherwise "reclaimed." Three-fourths of the 50 U.S. states lost considerable portions of their wetlands. Six states lost more than 85 percent, and another 22 states lost more than 50 percent. For over 200 years, "swamps" were considered to be barriers to progress or, worse, as sources of noxious vapors and pestilence.

Just what exactly is a "wetland"? The answer seems axiomatic at first glance: land that is inundated by water. As with most things, the definition is a bit more complicated. For example, one of the more popular college ecology textbooks, Robert Leo Smith's *Ecology and Field Biology*, lists more than 20 different kinds of wetlands, ranging from "seasonally flooded basins" (that actually are dry at certain times of the year) and "wooded swamps" to "freshwater marshes," "salt flats," and "prairie potholes." The U.S. Army Corps of Engineers defined wetlands in 1984 as:

> *"The term 'wetlands' means those areas that are inundated or saturated by surface or ground water at a frequency and duration sufficient to support, and under normal circumstances do support, a prevalence of*

*vegetation typically adapted for life in saturated soil conditions. Wetlands generally include swamps, marshes, bogs, and similar areas."*

The largest wetlands in the world are found in the poorly-drained permafrost areas of Siberia and Canada, and in the seasonally flooded lands of the Amazon River basin in South America. In the U.S., the Mississippi River floodplains represented the largest contiguous wetland area at one time.

During the 17th through 19th centuries "swampy wetlands" were drained for agricultural use. The muck lands left behind after drainage usually represented rich, fertile soils that were coveted by farmers. In some areas wetlands were drained to make way for roads and railroad rights-of-way. In 1849, the U.S. Congress authorized the State of Louisiana to "reclaim all swamps and overflow lands." By 1860, 14 other states followed suit and were involved in draining a total of 65 million acres of wetlands of all description. Forestry practices, such as clearcutting of the swamp forests of Ohio, Indiana and Illinois; river channeling and levee construction; and other flood control projects also destroyed wetlands at rapid rates.

Wetlands were often considered to be worthless, idle lands that needed to be put to use by a rapidly-growing country of hard-working and industrious people. Florida gubernatorial candidate, Napoleon Bonaparte Broward wrote in a campaign letter in 1904: "Today for hundreds of miles the mighty Mississippi River is confined by levees and so controlled that millions of acres of swamps have been turned into arable land and are in yearly cultivation, yet the water in the river is often 20 feet above the level of the farms. How much simpler would it be to drain the Florida Glades where no danger from overflow exists once canals of sufficient capacity were cut and the canals themselves would serve as effective channels of communication between different parts of the State." With hubris typical of the candidate, he added: "...shall the sovereign people of Florida supinely surrender to a few land pirates and purchased newspapers and supinely confess that they cannot knock a hole in a wall of coral and let a body of water obey a natural law and seek the level of the sea? To answer "yes" to such a question is to prove ourselves of freedom, happiness and prosperity." In an age of "can-do" attitude, Broward's words resonated with voters and he was elected governor in 1905. The governor kept his promise and during his term large portions of the Everglades were drained. As water levels dropped the sawgrass, and animals that depended on it, began to disappear.

Destructive practices like the draining and channeling of wetlands continued throughout many parts of North America well into the 1980s, despite accumulating bodies of evidence that wetlands had practical as well as intrinsic values. Voices of caution against the destruction were raised by biologists, conservationists, land managers and ordinary citizens.

In addition to the displacement and extirpation of numerous plant and wildlife species, loss of wetlands has been implicated in increased numbers of wildfires, disappearance of potable water sources, increased water pollution levels, disruption of natural systems of nutrient cycling, salt water intrusion into freshwater ecosystems, and even alteration of local climatic conditions.

The first conservation measures were actually begun in 1934, when the U.S. Congress passed the Migratory Bird Hunting Stamp Act, whose revenues were earmarked for wetlands purchase, protection and restoration. A gradual reversal of wetlands destruction practices emerged during the administration of President Franklin D. Roosevelt (FDR). Although different Presidents had differing levels of commitment for ecosystem protection – and lobbying efforts never stop to reverse the gains made in the last half century – the large-scale wanton destruction of wetlands is hopefully relegated to the pages of history books.

During every spring my wife and I join other bird watchers to spend two days at the Montezuma National Wildlife Refuge near Syracuse, New York. Our leader is an expert from the Cornell Lab of Ornithology and over a weekend's time we record 80-100 species of birds, as well as otter, muskrat, beavers and other animals that prefer to live in or around water. The highlight of our trip is seeing the roosting pairs of bald eagles caring for their hatchlings. If we are lucky, a panoramic sweep of the horizon can yield as many as 4-6 of these majestic birds in trees, on tops of electric towers or in mid-flight. I am by no means an expert birder and I feel fortunate if I can identify just a few avian species. But it doesn't matter; I enjoy the typically blustery early spring weather, the wind, the smells and sounds of the area, and the camaraderie of our group.

The Montezuma Refuge was established in 1938 as a resting and breeding grounds for migratory birds and other wildlife. It is located in one of the busiest corridors of the so-called Atlantic Flyway. The more than 7,000 acres of marshes, mudflats and shallow water habitats support about 240 different species of birds. On a good day, we see thousands of ducks, geese, sandpipers, terns, herons, plovers, snipes…and the bald eagles.

The history of Montezuma is typical of the life, near death, and rebirth of many wetlands in the United States. Originally, the area consisted of marsh and swamp land – remnants of the last ice age that also carved out the basins of the Finger Lakes of New York. Here, the native tribes hunted, fished and trapped aquatic mammals for their own use and for the fur trade. After the Indians were chased off their lands (well, there WERE a series of treaties but they often didn't mean much to the conquerors), the land underwent major changes.

In 1810, the New York State Legislature passed an act to drain the "Cayuga marshes." The Erie and the New York State Barge Canals were dug to facilitate the flow of goods and people from the "civilized" eastern parts of the state into the "wilderness." The widening of the barge canal in 1910, and the construction of a lock and dam on the north end of Cayuga Lake lowered water levels in the Seneca River by about 10 feet, and water gradually drained from the marshes. More drastic drainage ensued when the channels of the Seneca and Cayuga Rivers were straightened, and the dry land was planted for corn and other crops.

During the FDR Administration, 6,400 acres of the former marshland were acquired by the Federal government and the Civilian Conservation Corps began to rehabilitate the wetlands. The Montezuma Migratory Bird Refuge opened on September 12, 1938.

The latter half of the 20th century witnessed gradual wetlands restoration. Progress has been slow and hard-fought, but many former drainage projects, like Montezuma, and the Okefenokee Swamp straddling Georgia and northern Florida, have become National Wildlife Refuges. In 2011, there were 555 of these in the U.S.

One of the largest and costliest restoration projects involves the Everglades of Florida, which stretches for 100 miles from Lake Okeechobee to the Gulf Coast of that state. It is perhaps ironic that one of the most expensive drainage projects ever also became one of the most expensive on-going restoration efforts. Everglades National Park opened in 1947 during the Truman Administration and the battles and hard work of protection have continued since then. In the 1960s and '70s, for example, the city of Miami wanted to build the then largest airport in the world in the Everglades. Years of litigation followed before the project was finally abandoned. Some of the damage done to the Everglades is, of course, irreversible. Towns and highways, levees and farmland now occupy large tracts of the former wetlands.

Gone, too, are the swamp forests of Ohio and Indiana, and the cypress swamps of the Suwanee River basin in Georgia. But the gradual recognition that wetlands serve important wildlife and human functions has allowed for the rebirth of these natural areas. As the human population continues to grow, one of the most important problems is going to be procuring enough clean, potable water. Wetlands vegetation has been shown to play crucial roles in cleansing water of both natural and man-made pollutants. For example, experimental plantings of cattails and marsh grasses have yielded remarkable results in sopping up and removing from water many types of pollutants, including organic compounds, like benzene and toluene, and heavy metals, such as mercury. One of the best examples of this approach is the successful remediation of the toxic waste site of the former Sinclair/ARCO/BP oil refinery site in Wellsville, New York, which operated for over 70 years and polluted the Genesee River with petroleum by-products.

The story of wetlands is an important reminder that we humans are not always as smart as we think we are. Just because something becomes feasible from an engineering standpoint, it doesn't mean that it is beneficial in the long run either for people or for other living organisms.

Long live the wetlands!

## WHEN HEARING IS SEEING

*When I had written my 'Silverwing' series - I'd imbued*
*the bats with full human awareness and vocabulary.*
*-Kenneth Oppel*

Bam! Bam! Bam! The sounds of shotgun blasts in the twilight hours of a lazy summer evening were unmistakable. This wasn't a car backfiring. It was nowhere near hunting season and, although I lived in the country where discharging a firearm wasn't against the law, I was unaccustomed to hearing gunfire. Yes, we all owned some sort of a gun or another, but neither my neighbors nor I were trigger-happy people. So, this had to be something bad. All of this raced through my mind as I cast my book aside, scrambled from my hammock and ran toward my neighbor's house where the sound of shooting seemed to originate. I rounded the corner, breathless from the effort and fear.

My neighbor was on his porch, shotgun in hand and with an I-don't-know-whether-to-laugh-or-cry expression on his face. He had just blasted off the corner of his porch and was standing amidst a pile of splintered wood and roof tiles. Shaking sawdust from his hair and clothing, he explained somewhat sheepishly that he was trying to kill some of the dozens of bats flying erratically and taking shortcuts through his open porch in the waning daylight. He said that he had had enough of these bats "aiming for my wife's and daughter's hair." That, of course, was not the way the bats perceived the situation and that is the point of this story.

My neighbor's attitude toward bats is not unusual. Many people loathe these strange-looking "flying mice." Most of us have grown up with vampire and bat stories, and these creatures of the night evoke in us primitive fears and atavistic emotions associated with evil, witchcraft, sorcery, ghosts and goblins. Because of our childhood stories, we have come to detest the bats as much as we hate the "evil" wolves, snakes, toads and vultures.

The bat's unsavory reputation is totally undeserved. All of the bats inhabiting North America are small, mostly insect-eating mammals that should be considered friends rather than vicious foes. Many species of bats eat huge numbers of insect pests and mosquitoes, often consuming as much as half their own body weight in a single night's feeding.

Yet, it is easy to understand why there is so much gut-level animosity toward these animals. Just look at them. The bat is a relative of shrews and moles, but its body has been modified for flying and collecting its insect meal midair. The creature is soft and furry, but certainly not cuddly, with forelegs that have taken on the shape of wings. The bones of the four fingers of each front leg, but not the thumb, have elongated enormously to frame the support structure for the paper-thin smooth skin that forms the wings. The animal thus resembles a mythological creature – not quite mammal, not really a bird, but a freakish chimera of various odd body parts.

An up-close look at the head does nothing to dispel the bat's reputation for wickedness. The mouth is full of razor-sharp, chisel-shaped teeth that are perfect for crushing the hard-shelled bodies of insects. The face, ears and forehead are studded with grotesque, gargoyle-like enlargements that help guide the bat in tracking its prey. The ears are especially large and each ear has a well-developed attachment – the tragus – for collecting and funneling sound waves.

Capturing flying insects in the dark of night is no easy task. Bats have evolved a system of tracking, pursuing and capturing moving targets as small as a gnat or mosquito with remarkable accuracy. Although their vision is not as poor as the adage "blind as a bat" would suggest, eyesight alone is not very useful in the dark when bats are actively feeding. Instead, bats use an extremely sophisticated system of honing in on their prey.

In the 18th century, the great Italian scientist, Lazzaro Spallanzani observed that blinded bats were able to fly around obstacles. But bats whose ears were blocked or damaged could not navigate around various objects. It wasn't until the 1930s, though, that scientists discovered that insect-eating bats use an amazing sonar system for hunting. Science calls it echolocation. The bat emits a series of high-pitched sounds (ultrasonic pulses) which travel at high rates of speed and bounce off objects back toward the animal. The bat generates these sounds from the larynx (voice box) and the ultrasonic pulses are emitted either from the nostrils or via the lips. The tragus and other ear structures then can interpret the location of the target by the echo created as the pulse waves bounce off the object back to the bat's ears. Although humans can't hear them, these ultrasonic pulses may be emitted as frequently as 50 to 20,000 times per second, giving the bat an extremely accurate readout of the size, speed and direction of its prey.

The bat's sonar system is a highly advanced version of what modem submarines use for locating enemy vessels and seafloor features in the dark and murky depths of the ocean. In fact, the bat's echolocation system is more than sonar – it is more like the modern Doppler radar that can generate three-dimensional images of a moving object.

Several types of echo-detecting nerve cells are located in different areas of the bat's cerebral cortex. These neurons selectively respond to different types of stimuli and they, together with other areas of the brain and ears, allow for the sophisticated food-capturing mechanism to function in such a way that some bats can capture two mosquitoes per second for several hours of feeding time. (In an interesting example of co-evolution, some species of insects can detect that they have been "pinged" by the sonar of a bat, and they can either commence evasive maneuvers or try to jam the bat's echolocation system by emitting their own sonar signals. But that's another story.)

Just how well this system works for the bat has been amply demonstrated by experiments. In one series of laboratory trials, bats were allowed to fly in darkened rooms whose interiors were rigged with a system of thin wires spaced just far enough apart to allow the bat to pass through. In many replicated trials in which the wires were criss-crossed in random arrangements, the bats rarely made mistakes in interpreting the location of obstacles by getting tangled up in the wires. Ironically, this uncanny ability to judge distances and objects is also the reason that many people, like my neighbor, believe that the bats are "aiming for their hair." What may seem like a perfectly reasonable clearance for a bat may be an uncomfortably close encounter for a human.

Not all bats are insectivorous and not all bats use echolocation. Of the approximately 1,000 species of bats world-wide, many live in warm tropical environments where some species live on arthropods, lizards and even small rodents, while others are primarily fruit eaters. The latter group of bats can see well and they are mostly active during the daylight hours. These bats may do some damage to commercially grown fruit but their value as pollinators far outweighs any loss of crop.

Then there is the famous vampire bat which has inspired so much folklore and the Dracula stories. The vampire bat of the Family Phyllostomidae – or Desmodontidae, if you really want to split hairs! (sorry, I couldn't resist) – of Central and South America, is adapted to living on blood meals it gets from large herbivores, mostly horses and cattle. These bats make a small incision in the skin and, with the aid of an anticoagulant, lap the blood rather than suck it from their host. They occasionally even drink human blood. I once had a student from Guatemala who told me that she fell asleep on a beach only to wake up and find a vampire bat snacking on her.

In most of North America, we don't need to fear bats. There are occasional colonies where rabies becomes established, but a bat is no more likely to be the carrier of the rabies virus than a raccoon, fox or any other mammal.

My neighbor was a skilled builder and had the damage to his porch repaired in no time. He decided that shooting at the bats was not a good idea after all. And it was a waste of ammo. Since we lived in an area where old abandoned coal mines pockmarked the landscape, making ideal roosting sites for the bats who weren't going anywhere any time soon, there was nothing to do but to suggest that perhaps his wife and daughter visit a beauty parlor to have their hairdos rearranged.

# SHOULD WE STAY OR SHOULD WE GO?

*My youngest son and I are in the yard enjoying the warm October afternoon. Each of us has a bottle of Mr. Bubbles from which we whimsically create either one huge bubble or a bunch of small ones in rapid succession. The scattering of bubbles drives the cat crazy and she is running back and forth trying in a mostly futile effort to catch them. The sky is darkening from the west. Towering cumulus clouds gradually replace the afternoon sun and there is a smell of rain in the air.*

*I hear them long before I see them. Loud honking announces their imminent arrival and I look to the sky once again, this time with anticipation. An undulating and slightly crooked V-shaped formation of Canada geese appears over the treetops from the east. They are flying too high to make out their distinctive black-and-white faces and their chevroned tails, but their silhouettes are unmistakable even at this distance. There are*

*about 20 of them and we watch them until they*
*disappear from view.*

I wrote this entry in my journal many years ago. The boys are grown now and we don't blow bubbles when we get together. We talk about sports and jobs and houses and other such manly things. But we still stop the conversation when a flock of these birds traverses the sky overhead.

Canada geese (*Branta canadensis*) are ubiquitous in North America. This is reflected in the large numbers of common regional names given to these large handsome birds. In different parts of the continent they are known as the Canada goose, Canadian goose, French goose, Mexican goose, Eskimo goose, honker, bay goose, white chin, cravat goose, Canada brant, and at least half a dozen other vernaculars. Wildlife biologists often recognize different geographical races or subspecies based on body size and average weight, shapes of the bill and head, and assorted other physical features. Regardless of the subtle differences in appearance, all Canada geese are instantly recognizable by their long black necks and heads and white chin straps, soft grey breasts and black-and-white rump and tail feathers.

The North Atlantic populations of the Canada goose, designated as subspecies *canadensis* (someone lacked imagination, I guess), nest in the Canadian Maritime provinces and migrate in the fall to their wintering grounds in New England, the mid-Atlantic states, and as far south as the Carolinas. Other groups, however, embark on much longer semi-annual journeys: when population levels are high, some

34

members of the Alaskan subspecies travel as far as California and Texas to beat the winter blahs. According to K.H.Breen's lavishly illustrated book, *The Canada Goose*, the birds travel in "flocks consisting of extended family units." Most flocks fly at low altitudes (between 1,000 and 3,000 feet), but "they can easily fly over mountain ranges 12,000 feet high."

Winter quarter requirements include open waters, which the birds not only use for food but also as their preferred sleeping place. An ornithologist acquaintance told me that "the geese use fresh water lakes and ponds and salt marshes in the winter where their diet consists of a variety of marsh grasses and invertebrates. When these waters freeze over and are snow-covered, the birds often head for the coast," where there is more likely to be open water.

There seem to be two very distinct behavioral groups of Canada geese in a given locality in the wintertime. The first group consists of migrants who faithfully return, year after year, from their summer ranges. The second group is made up of birds who are year-around residents and don't migrate at all. For whatever reason, these birds have lost the compulsion to take to the skies with the arrival of spring and live, rear their young, and overwinter in the same locale. Lots of theories and conjecture abound as to why these different behaviors are seen among different clans of birds, but the ones who stay put often become quite a nuisance in urban areas. Wildlife biologists estimate that the migratory populations have actually declined in numbers over the last two decades, while non-migrant numbers have steadily increased.

Another journal entry:

*My friends and I are carrying blankets and picnic lunches, looking for a suitable spot in the grass to spend the lunch hour. This riverfront park is one of our favorite places to bask in the sun and watch the kayaks and the scullers plying the shimmering water of the Schuylkill River.*

*As we stretch out on the blanket and unpack the food, we attract the attention of a flock of resident Canada geese grazing nearby. One by one, they waddle closer and soon we are surrounded by 20-30 birds. They are huge and imposing at eye level. They are looking for a handout and we begin to feel like hostages as they tighten the circle around us. Hitchcock's birds come to mind.*

*Suddenly and completely unexpectedly the entire entourage is alerted by a posted sentry and they take to flight. Huge bodies require a running start and they lift off without a trace of grace. Some of them clear our heads by only a few inches as they gain altitude. This time, I can't say that I'm sorry to see them go.*

Aside from snide remarks about how they are becoming lazy like modern humans, a number of possible explanations for this shift in migratory behavior have been put forth. One hypothesis is that at one time people kept many of the birds as live decoys. When this practice became illegal, the geese were simply turned loose but they were already "set in their ways" and didn't fly away.

A more likely explanation is that an abundance of food may have led to the establishment of sedentary populations. Golf courses, man-made ponds, corn stubble, conservation areas, protected wetlands, and various landscape practices also may have provided safe havens for the birds.

> *Many years ago there was a widely quoted anecdote about the then mayor of Philadelphia, a man named Frank Rizzo. He was a no-nonsense guy whom many people loved because he got results. After several news stories aired on local TV stations detailing the plight of a flock of resident Canada geese along the banks of the Schuylkill River during an especially harsh winter, Rizzo was reported to have become agitated about the numbers of phone calls received by his office to do something. His Honor is reported to have called his Parks and Recreation commissioner and said: "I don't care what it takes, just feed those (expletive deleted) ducks." The birds had food within hours.*

The geese also seem to have learned to frequent areas where hunting is prohibited. Robert Elman notes in his *Hunters Field Guide* that "the extreme intelligence of

Canadas shows in their preference for those refuge waters where shooting is illegal."

Undeniably, Canada geese have become a major problem in some areas. In addition to their excrement, the birds tend to interfere with vehicular traffic in and around shopping malls, parks and other recreation areas. They frequent golf courses where they not only obstruct the greens but also ruin tender young grass. I talked with a golf course manager once about the birds and his only solution was to "keep them harassed." The man donned a bright red jacket and drove his golf cart right at the flock at maximum speed. After a while, it was enough for him to just wear the red jacket. The birds took off, at least temporarily, as soon as they spotted him standing there.

# *QUO VADIS, BIRD FLU?*

*O'er ladies' lips, who straight on kisses dream,*
*Which oft the angry Mab with blisters plagues'*
*Because their breaths with sweetness tainted are.*
*-William Shakespeare*

Several years ago I gave a talk for a group of college students on bird flu. At that time, in 2005-2006, there was very great concern about the possibility that an outbreak of a deadly influenza pandemic was upon the world. Experts likened this virus to the "Spanish flu" that killed so many people near the end of World War I. The apocalyptic event predicted in 2006 never materialized and headlines and sound bites slowly faded and then disappeared from newspapers, magazines and television. However, bird flu was back in the news in the spring of 2012, but this time around for quite a different reason.

The origin of the word "influenza" ("flu" for short) is rooted in a Latin word because at one time it was believed that the influence of the stars was responsible for widespread outbreaks of the disease in human populations. Epidemics have been described since the time of Hippocrates in the 5th century, B.C., but the causal agents – which can be either bacteria or viruses – were not known until recently.

As for viral influenza, there are three main flu virus groups, known as A, B, and C. Most cases of respiratory illness and death are caused by influenza virus A. Within this group there are many "strains" or "subtypes," representing mutations in the genetics of the viral particle. Since viruses are so simple, even minor alterations in their genetic makeup can lead to major changes in their structures and virulence. The most common mutations typically affect one or both of two easily detectable parts of the virus: the so-called hemagglutinin (H) part and/or the neuraminidase (N) component. Without going into extended detail, the changes in these proteins are designated with numbers. Thus, one can speak of H3N2, commonly known as the "swine flu from Taiwan in 1970," or the H7N7 strain, found in 1956 in horses. The infamous "Spanish flu" pandemic of 1918-1919 has been identified as H1N1, while the most recent "bird flu" virus is designated as H5N1.

Viral infection often confers immunity for the host affected because the body makes powerful proteins, called antibodies against the invader. The antibodies have a chemical "memory" and the next time the virus attacks the body, it is quickly immobilized. If, however, the virus changes, the host doesn't recognize it as a previously encountered adversary and new antibodies must be made. The highly mutable nature of the flu virus is the reason that yearly flu vaccinations are necessary.

The moniker "Spanish flu" gives the mistaken impression that the influenza pandemic started in that country. The origin of H1N1 is unclear – some say it came from China; others point to its possible start in a military camp in Kansas in the United States. From wherever it came, it quickly spread, facilitated by troop movements during World War I and disseminated through commercial and military shipping lanes. World-wide estimates of mortality vary widely between 20-50 million victims during 1918-1919. Some estimates go even higher, possibly as high as 100 million people succumbing in a short period of time. In Spain, an estimated 8 million people died. In India, as many as 20 million souls may have perished, but many cases probably went unreported. One-fourth of the U.S. population were affected by flu and upward of 650,000 died. There were 50,000 deaths in Canada. No country went unscathed and there was no cure, no vaccine, no useful treatment.

Interestingly, young people were more likely to die than the aged. This observation went unexplained until recently, when it was suggested that the strong immune response of the young was actually their undoing: the violent reaction of the body was the cause of death, rather than the actual infection itself. What one physician of the time described as "[patients] died struggling to clear their airway of a blood-tinged froth that sometimes gushed from their nose and mouth" was actually a reaction of the body to the virus. The overall mortality rate in the Spanish flu outbreak was estimated to be 2.5 percent of people infected. In comparison, the "typical" annual mortality rate from the common viral flu is about 0.1 percent, mostly consisting of old and infirm individuals.

The influenza virus is ubiquitous, unpredictable and persistent. Sometimes, it is found only in certain localized areas, where it is established as an endemic agent of disease.

Other times, it spreads to areas outside its native territory and causes widespread epidemics in a population. When it crosses continents and becomes distributed world-wide, epidemiologists talk about pandemics. There have been numerous pandemics in recent memory, such as the "Asian flu" in 1957, the "Hong Kong flu" in 1968, or the "swine flu" of the 1980s.

Different varieties of the flu virus have different animal targets. Some strains can affect a large number of different hosts, including humans. In 2003, a highly virulent form of the influenza A virus was found among birds in southeast Asia. Initially found in poultry in Korea, Vietnam, Thailand, Cambodia, Laos, China, Japan, and Indonesia, it quickly spread among birds in the rest of Asia and most of Europe.

Identified as strain H5N1, it also was responsible for human illness in the aforementioned countries, as well as in Azerbaijan, Turkey, and Iraq. To arrest the spread of the disease, several hundred millions of chickens, geese and ducks were destroyed and buried, but migratory wild birds continued to disseminate the virus to new areas. Only a few hundred people became infected, and many of those survived. Nevertheless, in 2006, experts determined that H5N1 had the "genetic signature" of the 1918 strain and that it could become a pandemic with apocalyptic consequences.

That particular prediction did not materialize for several reasons, among them the quick destruction of infected poultry, a better understanding of how the virus spreads from bird to humans, massive education campaigns, and avoidance and quarantine measures. But the most important reason was that, unlike its cousin in 1918, H5N1 lacked a good mechanism to spread from one human to another. And that's where the current news about the bird flu virus comes in.

In order to become efficient at spreading through the human population, H5N1 needed certain genetic traits. Virologists compared the genomes of H1N1 and H5N1 and predicted that three of 8 genes needed a total of about 20 mutations in the bird flu virus in order to become fully adapted to the human host. While that sounds like a lot of required changes, for viruses that is an easily doable task. In fact, by 2006, four of the 20 mutations had naturally occurred in the viral population.

Researchers were interested in the mechanisms of the mutations and began experimenting with conversion of the virus to become a human pathogen. These experiments were conducted in reputable laboratories by serious molecular biologists in Europe and in the United States. Several virologists had been able to tinker with the viral genome and announced in early 2012 that they succeeded in achieving the required mutations to make this virus a human pathogen.

The rationale for this type of research was that by understanding the mechanisms and paths of mutations, the potential for spread by this and other viruses among human populations in the future could become more predictable. But, therein lies the rub: by creating a highly virulent strain in the laboratory, there is opportunity for misuse, as well. These so-called "dual purpose" experiments mean that the purported benefits of research may be counterbalanced by their potential application as biological weapons of mass destruction.

Researchers ask questions, solve riddles, and usually share the results of their work with the rest of the scientific community. The questions are these: Should details of these experiments be published in journals where anyone could gain access to the information for good or ill purposes? Does the right of the scientist to reveal the methods and results of research without fear of censorship outweigh the need for biosecurity?

These are issues in which bioethicists, researchers, institutional committees, government officials and an educated public must make their voices heard. This last group is especially important, but that also means that every-day citizens must be knowledgeable and pay attention to policies. Otherwise, the old saying may come true: "The road to hell is paved with good intentions."

# "THE CELL FROM HELL"

*One may not doubt that, somehow, good*
*Shall come of water and mud;*
*And, sure, the reverent eye must see*
*A purpose in liquidity.*
*-Rupert Brooke*

What can cause millions of dollars of damage without even trying? What can lead to confusion, memory loss, rashes and a host of other human maladies? What can reportedly take on dozens of different forms, shapes and modes of existence? What had become Public Enemy Number 1, at least for a short time, in the U.S. in the 1990s? Why, *Pfiesteria*, of course!

In the late 1980s, researchers at North Carolina State University reported on the discovery of a new microscopic organism. The creature belongs to a group of one-celled microbes called dinoflagellates. These tiny aquatic life forms have a hard cellulose and glass cell wall, typically two whip-like structures, called flagella, which they use for navigation, and a "true" nucleus. Most people hear of them only when toxic "red tides" occur near the edges of oceans, killing fish and closing beaches until the population of these dinoflagellates decline. Most marine biologists, in fact, have a much kinder view of them and, collectively, all the different species are believed to be the most important

45

sources of nourishment in the food chain for all ocean-dwelling animals.

The dinoflagellates are a curious lot: they straddle the line between animal-like characteristics (motility, feeding, and irritability) and plant-like features (presence of chloroplast and cellulose). Their fossils date back to almost 600 million years ago.

The accidental discovery of a new dinoflagellate began in 1988 when fish added to a tank containing Pamlico River, North Carolina, water showed symptoms of poisoning and began to die. Detective work by Drs. Edward Noga and JoAnn Burkholder pointed to a novel microscopic culprit, which they named *Pfiesteria piscicida.* Fish killer!

Intensive laboratory and field investigations began during which several lab workers became ill with symptoms of memory loss, fatigue, confusion, skin rashes, and other physical problems. These symptoms were thought to be associated with toxins produced by *Pfiesteria.* There was just one problem: typical toxins associated with dinoflagellates could not be found in the air, water, or laboratory equipment.

The more the researchers looked, the more life stages of *Pfiesteria* there seemed to be: immobile cyst forms, amoeboid creeping forms, swimming zoospores, sexual and asexual stages seemed to abound. In all, 24 different forms were described and a map of the life cycles looked like a person suffering from *petit mal* seizures trying to draw lines on a piece of paper.

Needless to say, the scientific community was skeptical of these findings. I remember seeing articles appearing in scientific journals in which various authorities argued about the possibility of such a large number of chameleon-like forms. From an evolutionary point, having 24 different morphological types makes little sense. I know of fungi, called "rusts" that have as many as 5 spore stages and two different host plants. Even those numbers represent a tremendous energy investment in survival and propagation. But 24 different forms?

Labs trying to replicate the results met with little success. Sexually reproducing forms never were found; most of the different life forms were pronounced to be artifacts or altogether other creatures. As the scientific community debated, tempers flared. Negative comments about laboratory methodology, statistical evaluation, and data interpretation degenerated into personal attacks and questions of objectivity. There was even hushed talk that the *Pfiesteria* work was attacked not on scientific principles but because one of the principal investigators was a woman and science, after all, is still a domain of men.

All of this squabbling may have stayed within the scientific community had it not been for successive major fish kills in 1997 in the Pamlico-Albemarle estuaries of coastal North Carolina, and in the Chesapeake Bay in Maryland. In what can only be described as the "perfect storm" of economic losses from dead fish, media-fueled public hysteria, and people's fascination with the image of a mysterious, invisible, chameleon-like creature, *Pfiesteria* was catapulted into the headlines for months.

In 1997-98, *The Washington Post* published 130 news articles related to the organism. Not to be outdone, *The Baltimore Sun* ran 170 stories, and the microbe made the ABC, NBC, and CBS nightly national news. By the time the dust settled, the fisheries of Maryland and nearby states lost over $50 million in revenue to the publicity caused by the "fish killer," "ambush predator," "fish AIDS," or the "cell from hell," as the colorful headlines referred to it.

To be sure, fish kill and *Pfiesteria* were never directly linked by epidemiological data. All of the "evidence" consisted of supposition, conjecture and unscientific guesswork. Yes, *Pfiesteria*, like many other microorganisms, including other dinoflagellates, can generate toxic byproducts when populations are high. That statement also could be made for cows, cats, and humans, as well. For *Pfiesteria piscicida*, at least, it seems that high phosphate levels in water can cause a population explosion, triggering production of certain toxic free-radicals which can cause skin lesions in fish, possibly leading to infection caused by opportunistic bacteria and fungi present in the water.

In my environmental science class I often talk about the interconnectedness of all parts of an ecosystem. Whenever the delicate balance among organisms and their environment is disturbed, bad things can happen. If there is a take-home message in the *Pfiesteria* story, it is that instead of looking for a single menacing threat we should focus on maintaining or restoring the proper ecological balance in the environment. That includes keeping nutrients, like phosphorus and nitrogen, out of the water, and generally treating our most precious commodity, water, with the respect it deserves.

# "EVERY FACE GATHERED

# PALENESS..."

*Our ancestors suffered – all of them, including kings and
queens, industrial magnates and country squires, for wealth
brought little respite when even the best doctors
could do next to nothing.*
*- Stephen Jay Gould*


OK, *Pfiesteria,* move over! You are not such a tough
guy, after all. In the annals of waterborne diseases, there are
much more serious and widespread organisms than the
protozoan parasite chronicled in the previous story.
Interestingly enough, these big-time players don't seem to
get the publicity – hence the notoriety – that *Pfiesteria* did a
few years ago. Although we tend to think of these diseases in
historical terms, they are still around today in many parts of
the world.

In April, 1993, nearly 40 percent of the more than
850,000 people in Milwaukee, Wisconsin, served by the
city's water treatment plant became ill with symptoms of
fever, cramps and severe diarrhea. A total of 104 deaths,
$31.7 million in medical costs, and an estimated loss of
productivity of $64.5 million were attributed to "Crypto," an
intestinal parasite. The organism, *Cryptosporidium parvum,*
is "one of the most frequent causes of waterborne diseases

among humans in the United States," according to the Centers of Disease Control and Prevention (CDC&P). For 2011, the agency estimated 750,000 cases of illness due to this critter. Crypto may be one of the most frequent causes of what most people mistakenly call "the stomach flu" or the "24-hour bug." It is distributed world-wide and, in addition to unclean water, it also can be spread from hand-to-mouth, person to person transmission, uncooked food, or certain sexual practices.

*Cryptosporidium* is a one-celled parasite that lives in the intestines of mammals, including humans. It is very resistant to chlorine, the chemical most commonly used to disinfect water. In the case of the Wisconsin outbreak, the pathogen was suspected to have washed into the Milwaukee River from nearby upstream farms from livestock feces carried by precipitation and snowmelt.

Such large scale outbreaks of waterborne diseases are not common today in the industrialized world. One of the greatest advances of $20^{th}$ century science and engineering technology has been to insure the safety and cleanliness of municipal water supplies through sanitation, water purification and education. This was not always the case throughout recorded history.

Europe, Asia, and the new world colonies were repeatedly visited by catastrophic water-borne epidemics during the past thousand years. Pestilence of every form struck terror in the hearts of city dwellers and country folk alike, causing great suffering and social upheaval. Many of these illnesses spread like wildfire and were magnified by the crowded, unsanitary conditions which prevailed among inhabitants of cities small and large. Causal organisms of typhoid, cholera, amoebic dysentery, hepatitis A, giardiasis and a number of parasitic worms and flukes were, at times, present in the water supply. But, of course, no one knew anything about microscopic organisms as causal agents of disease until the 1870s.

Severe, and often fatal diarrhea caused by Asiatic cholera was first recorded in India in 1816. It took less than a decade for this bacterium to spread to most of Southeast Asia via the traditional trade routes. The principal means of transmission was by fecal contamination of water. By the 1830s, epidemic proportions of cholera were found in Russia, Germany and England, where ballast water from ships coming from India was suspected to be the initial source of contamination.

Between 1832 and 1900, waves of cholera epidemics struck parts of Asia, Europe and North America. According to contemporary accounts, in 1832, "fifty-one thousand-seven hundred immigrants from England and Ireland arrived at the port of Quebec, and in every city whence these immigrants came, cholera was epidemic." Soon, the disease appeared in New York City, where more than 3,500 people died, and about a third of the 70,000 inhabitants fled to escape the pestilence.

Between 1847-48, more than a million people died from cholera in Russia, China, other parts of Asia, and in Europe.

In 1848 cholera killed more than 6,000 in Cincinnati, and a newspaper correspondent traveling in Sacramento, California, reported: "We...visited the burial ground where I saw the long parallel lines of graves of cholera victims...resting places of upward of 1,700 people who had fallen in 15 months." In the 1850s, Canadian authorities tried in vain to halt the advance of the disease, going so far as to install "a battery of two 12-, and one 18-pound guns in the center of the [St. Lawrence] river" to stop and quarantine ships arriving from Europe. This attempt failed miserably because the nature of the illness and the mode of transmission were not understood.

Conditions were especially bad in military encampments and in cities under siege during wartime. The crowded, unsanitary conditions in military camps, open latrines used by soldiers, lack of provisions and clean potable water, along with the physical and mental stresses of forced marches, fatigue and fear, were ideal breeding ground for disease.

In what was a prelude to the "French and Indian Wars," New England troops besieged the French garrison of Louisburg, Nova Scotia, in 1745. After the fall of fortifications, one observer wrote of conditions in the town: "Since we besieged – there was such numbers kill'd and that died – with a sickness that they digg'd a hole about 12 feet square and about as deep where they threw in all together and without coffins I was told." Although the cause(s) of the conditions were not known, one Massachusetts physician accompanying the troops commented that "the town had become so filthy that people died like rotten sheep."

General R.E. Lee wrote a letter to his wife during the early stages of the American Civil War: "Sept. 3, 1861: Rain, rain, rain, there has been nothing but rain. This state of weather has aggravated the sickness that has attacked the whole army... diarrhea, measles and typhoid fever... This makes a terrible hole in our effectiveness." During the four years of the conflict, more soldiers and civilians died from diseases, including cholera, than from injuries incurred during battles.

Cholera got its name from the Greek word "chole", meaning bile, as at one time excessive production of that substance was believed to be the cause. John Snow, the "Father of Epidemiology" clearly showed the relationship in his famous London Broad Street pump experiment between contaminated water and the life-threatening diarrheal illness. Although Filippo Pacini is said to have discovered the bacterium, *Vibrio cholerae*, in 1854, it was the German physician, Robert Koch, who is often credited with its identification. Certainly, it was Koch who showed the cause and effect relationship between the organism and the disease. Koch, along with Pasteur, Lister, Semmelweis, and other scientists of the late 19th century is associated with the modern science of microbiology and the germ theory of disease.

However, it long remained a mystery how the bacterium survives without a mammalian host for lengthy periods of time between epidemics. Recently, researchers established the complex relationship between certain environmental factors, aquatic microorganisms, filter feeders (like crabs and oysters) as reservoirs of the bacterium in contaminated water, and the onset of human epidemics from tainted food or water.

Cholera is still a deadly disease today in many parts of the world, as the recent epidemic in Haiti, following a major earthquake and the subsequent unsanitary conditions due to destruction of municipal infrastructure, demonstrates. World-wide, cholera is estimated to affect 3.5 million people, with about 100,000 fatalities annually, due to lack of clean water and food.

Malaria and yellow fever, although not directly dependent on contaminated water for their transmission from person to person, were nevertheless associated with water because the mosquitos that carry the causal agents of these diseases require shallow, stagnant bodies of water for their reproduction. In Havana, Cuba, an estimated one-third of the population died from yellow fever in 1649. In that terrible summer of 1699, when yellow fever struck Philadelphia, Thomas Story wrote in his diary: "Great was the fear that fell upon all flesh; I saw no lofty or airy countenances…but every face gathered paleness, and many hearts were humbled…There is not a day and night that passed for several weeks [without] accounts of death or sickness of some friend or neighbor." Five members of Story's family died that summer.

No one understood the cause of the illness and death. Physicians could do little but comfort the sick and dying. Benjamin Rush, the famous Philadelphia surgeon, went so far in a desperate attempt to ward off the "miasma," that he suggested exploding gunpowder in dwellings. Needless to say, many people who tried that approach were injured by flying debris and shrapnel without any effect on the disease. For the next two centuries the city's population was repeatedly decimated by the viral disease. Memphis, New Orleans, New York and many other cities fared no better.

During the Spanish-American War of 1898, fewer than 400 American troops died in combat but the War Department reported 2,000 fatalities due to yellow fever. Lt. Col. Theodore Roosevelt wrote to Secretary of War, Russell Alger in June from the road to Santiago, Cuba: "If we are kept here it will in all human possibility mean an appalling disaster, for the surgeons estimate that over half the army, if kept here during the sickly season, will die." The government finally appointed a Yellow Fever Board, and Major Walter Reed's team solved the riddle of the virus and its agent of transmission, the mosquito.

Today's water supplies are not immune from contamination and disease, as the Milwaukee and Haiti examples show. World-wide, the biggest challenge is to provide safe, potable water for growing human populations. The World Health Organization estimates that 30-40 percent of the planet's inhabitants don't have access to this most important commodity.

Industrialized nations face other threats to their water supplies: chemical contamination of groundwater from industrial, agricultural and consumer sources, or from aging and neglected water systems, as the recent example of Flint,

Michigan, has shown. Some of these chemicals are not filtered out effectively by the natural filtering action of the soil, sand and clay particles that are natural cleansing agents of water.

Malaria continues to be the Number One cause of illness and death worldwide. In many countries, the protozoan causal agents of malaria have become resistant or immune to the most readily available antimalarial drugs. We have much more work to do! But the first thing is to "respect" microorganisms for what they are: extremely successful, extremely resilient, and ever-present life forms.

# A DIFFERENT KIND OF BANK

*The pastures are clothed with flocks; the valleys also are*
*covered over with corn; they shout for joy, they also sing.*
*-Psalm 65:13*

People who are knowledgeable about such things estimate that multicellular photosynthetic organisms – the plants – made it onto land during the Silurian Period in geologic history. During the past 450 million years or so, terrestrial plants diversified into more than 300,000 species. Of these, the most successful group today is the flowering plants, which account for about 275,000 known species. Only a small fraction, about 7,000 species, are used for agricultural purposes and of these, 30 make up most of the world's staple food sources. Wheat, corn and rice account for more than half of the daily calories humans consume today.

At some point during the evolution of human societies, someone somewhere realized that discarded seeds sometimes sprouted and could provide a staple food source without having to move onto new areas in search of food. Where agricultural settlements began in human history is not entirely clear but anthropological data point to the "Fertile Crescent" of the Near East as the center of wheat- and barley-based settlements about 10,500 years ago.

58

A few years ago there was quite an excitement (well, at least among those who get excited by such things!), when Israeli scientists discovered what appeared to be carbonized remains of domesticated figs in the Lower Jordan Valley, near Jericho. Although the findings were not universally accepted, if true, these artifacts push the evidence for purposeful planting of crop plants about 1,000 years before that in the Tigris-Euphrates Rivers region.

Archeologists have proposed many centers of origin for agriculture around the world at different times. In addition to the Near East, Northern China is thought to have the second oldest center at 9,000 years ago, followed by Central Mexico (5,700 years ago), the Andes region of South America around 5,200 years ago, and West Africa and North America about 4,500 and 4,000 years ago, respectively.

Two basic types of agricultural systems have been proposed: seed crops, such as wheat, maize, rice, barley, etc., which are labor-intensive and more vulnerable to natural disasters (e.g. hail, locusts, floods) that lead to crop failure; and "vegeculture," in which root and tree crops, like avocado, manioc, yam and potato are grown. These require less manual labor and are less subjected to large-scale destruction by natural forces, but the caloric value of the foods they produce is also usually less.

Scholars sometimes argue over what the first crops were and where they were first domesticated. With the discovery of the fig remains, archeologist Ofer Bar-Yosef was quoted in a June 2, 2006, *New York Times* article: "Eleven thousand years ago, there was a critical switch in the human mind – from exploiting the earth as it is, to actively changing the environment to suit our needs. People decided to intervene in nature and supply their own food, rather than relying on what was provided by the gods." In other words, people learned to exploit plants.

But there is another way to look at this. In a beautifully written book, *The Botany of Desire*, author Michael Pollan argues that plants use us as much as we exploit them in "a clever evolutionary strategy for advancing their own interests." Those interests include protection and dispersal by humans.

Whoever exploited who is a matter for semantics debate. Just the same, some people are interested in the origins of crops for more practical than philosophical reasons. One of the individuals who had done an extraordinary amount of work on the origins of cultivated plants was the Russian scientist, Nikolai Ivanovich Vavilov. Inspired by the writings of the Swiss botanist, Alphonse DeCandolle, on plant geography, Vavilov spent most of his scientific life traveling around the world, collecting, saving, and cataloguing seeds and other plant specimens.

Whereas DeCandolle, son of a protestant minister and a devout Creationist, believed that each new species arose from a specially "created" individual, Vavilov embraced Darwinian evolutionary concepts. Having spent some time in the laboratory of the renowned British biologist, William Bateson, Vavilov was well-versed in the modern 20th century science of evolutionary genetics and eventually mapped eight centers of origin for domesticated grain crops. He proposed both primary and secondary centers of origin for many important cereals. In a 1926 essay, for example, he writes: "...bottom cultivated wheat has two centers of origin: North Africa and southwest Asia...during very ancient times Egypt and other countries in North Africa had developed their own groups of cultivated wheat, sharply isolated from Asiatic wheat."

Vavilov could make such claims because he actually traveled to these places to examine not only the cultivated varieties but also their "wild" relatives. And he made another valuable contribution: he proposed that the centers of origin should be the sources of greatest genetic variability. He suggested that many wild relatives of modern crops should be able to be located in mountainous areas, as these would offer the greatest degree of environmental variation and, thus, harbor the greatest genetic diversity in the population. This "gene pool" could then be the source of new, improved varieties of crops.

It is mind-boggling to think that Vavilov eventually died of starvation and neglect in one of Stalin's famously brutal prisons. Vavilov's "crime" was that he disagreed with one of Stalin's protégés, a man named Lysenko. Lysenko almost single-handedly ruined Soviet agriculture with his erroneous concepts of crop genetics, but he was an "insider" in the dictator's inner circle.

As a child growing up in Communist Hungary, I frequently heard the name Lysenko and that of another favorite agronomist, Michurin. In the 1950s and '60s, Vavilov's name was never mentioned in Hungary in a carefully regulated and highly controlled system of Soviet-era propaganda and misinformation.

Modern crop science is a multibillion dollar business. Biotechnology promises to revolutionize agriculture, but such claims have been made before, as was the case with the "Green Revolution." Traditional breeding programs may be slower, but they rely on the more important principle of genetic diversity as the source of variation. For example, rice is not one uniform crop. There are roughly 40,000 different varieties of rice, belonging to four main categories: indica, japonica, aromatic, and glutinous. Each variety has slightly (or vastly) different genetic characteristics. In addition to nutritional value, taste, aroma, and stickiness upon cooking, there are also variations in yield, disease resistance, and ease of cultivation. These variations represent "insurance policies" because monoculture may lead to complete crop failure as the result of a single environmental change.

Vavilov recognized the importance of these genetic variations and spent three decades identifying, collecting and storing seeds from all over the world in his Institute of Plant industry in St. Petersburg, Russia. He looked at his collection as a savings account, a bank deposit of genetic diversity to be drawn upon in the development of new crop varieties.

The idea caught on and there are about 1,400 seed banks around the world today. The three largest ones include the Millennium Seed Bank in London, the Svalbard Global Seed

Bank in Norway, and Vavilov's seed bank in Russia. Other smaller, more specialized centers are the International Center for Tropical Agriculture (cassava, beans) in Coli, Colombia; the International Potato Center in Lima, Peru; the International Institute for Tropical Agriculture (cowpea, yams, soybeans) in Nigeria; and The International Rice Research Institute in the Philippines, which houses about 100,000 rice samples in its gene bank collection.

The Millennium Seed Bank in England is by far the largest repository of plant samples, consisting of about 100 times more specimens than the one in Norway. The Svalbard Seed Bank opened in 2008 in a converted old mine near the Arctic Circle. A joint venture between the Global Crop Diversity Trust, the Consultive Group on International Agriculture, the Norwegian government, and private donors, 400,000 seed samples, representing about one-third of the world's most important food crop varieties, are housed in a climate-controlled, earthquake- and nuclear bomb-proof facility.

Seeds are classified as "orthodox" or "recalcitrant." Orthodox seeds can stay viable for very long periods of time. For example, date palm seeds as old as 2,000 years have been successfully germinated and there is a recent report from Russia that 32,000 year-old seeds from the permafrost of Siberia have been found to be viable. Recalcitrant seeds cannot be maintained for long periods and must be continuously grown and harvested.

Either way, these so-called gene banks represent food security for a hungry world. These natural sources of variation are likely to represent the most important savings accounts for the foreseeable future. The seed banks guard something far more precious than gold or money: they are the depositories of the sustenance on which humans and other animals depend. Their importance is likely to become magnified as the human population continues its exponential growth.

# ROUNDWORMS GALORE

*Well do I remember the first tomato I ever*
*saw. I was ten years old, and was running*
*down one of those old-fashioned lanes, on*
*either side of which was a high rail fence,*
*then so familiar to all Ohio people.*
*-A.W. Livingston, 1893*

It's the middle of February. I look out the window of my study at a bleak winter scene. No, that's not right – the snow-covered trees are quite pretty and there are dozens of blue jays and snowbirds at the feeders. What makes it bleak is that the falling snow is blowing horizontally, driven by a wicked westerly wind.

But inside the house it's cozy and I am surrounded by a dozen seed catalogues. The seed supply companies are clever. They send their brochures in January and February to entice us northerners suffering from cabin fever to look at hundreds of photos of luscious fruits, vegetables and flowers. The seed sellers are whetting our appetites for the coming growing season. They know that we are probably going to get carried away and order too many kinds of seeds, bulbs, corms, canes, rhizomes, and bedding plants in our enthusiastic zeal to get out there, dig the soil, and plant our gardens.

I am looking at pictures of dozens of tomato varieties. Many have creative names, like "Early Girl," "Big Boy," "Fourth of July," etc. The names are followed by letters like V, F, N, T, A and others. Each letter stands for resistance of the tomato cultivar to a particular pathogen or pest. For example, "V" is for *Verticillium*, a fungus that causes wilting, loss of crop, or death of the plant. "A" and "F" stand for two other fungi, *Alternaria* and *Fusarium*, which can infect plants and ruin harvests. But right now, I am interested in the letter "N." It stands for nematode.

Most nematodes are microscopic-sized roundworms. Most are free-living, meaning that they don't parasitize other organisms or cause disease. They are present just about everywhere on Earth on land and in water, except perhaps in the Arctic or Antarctic ice sheets. A spoonful of soil may harbor hundreds of thousands or more of these little worms. In the soil, they eat bacteria, plant debris, and other detritus and they are not harmful to plants, humans or other animals.

There are some nasty species of nematodes. In animal hosts, they may be causal agents of trichinosis, Guinea worm disease, hookworm infections, heartworm of dogs, and a handful of other disgusting and debilitating causes of parasitism. In underdeveloped countries without clean water or sanitation, they are responsible for billions of cases of human disease, suffering, or death.

In plants, one of the most widespread and destructive roundworms is the root-knot nematode, *Meloidogyne*. The name of the genus is derived from the Greek words "meelo" (apple) and "gyne" (female), because the egg-bearing female that causes plant infections becomes a sedentary, swollen creature that is shaped somewhat like an apple. Like all other nematodes, this is a non-segmented roundworm (unlike the earthworm, whose body has dozens of clearly distinguishable segments). There are about one hundred different species of *Meloidogyne* world-wide and they attack the roots of more than one thousand different kinds of plants. In addition to the root-knot nematode there are a handful of other plant parasitic roundworms. Collectively, these nematodes are responsible for about 14 percent of all crop failures, and an estimated $100 billion in agricultural losses world-wide annually. By far, most of these staggering losses are due to the parasitic activities of the various species of *Meloidogyne*.

Root-knot nematodes have complex life cycles, during which they undergo several juvenile stages before becoming adult worms. What fascinates me is that females can reproduce with or without a male's contribution of chromosomes. In fact, most of the time males are not used at all in producing offspring.

Nematode females are not the only animals to make babies without a male. The process, called parthenogenesis ("virgin creation"), is well-known among insects, amphibians, sharks and other fishes, and even some bird species.

In populations of *Meloidogyne* males are rarely encountered. During "good times" (i.e., plenty of food, adequate soil moisture, etc.), females can rapidly produce baby worms to colonize available hosts and exclude competitors without having to first find a male, copulate, and then tend to fertilized eggs. The potential drawback is that all of the offspring are genetically identical to each other and their mother.

However, as it is the case with many insect species, males may be produced in large numbers during periods of extreme environmental stress. Difficult living conditions seem to trigger chemical changes in unfertilized eggs, which normally all become females, to also produce male worms. It's not that female parents need strong, virile blokes to get them through hard times. Rather, the male's contribution of chromosomes to the babies allows for greater genetic diversity in the population. This diversity is the insurance policy that potentially allows at least some members of the population to survive and cope with a changed environment.

While parthenogenesis is not known to occur among mammals, it has been intensely studied in other organisms to better understand the development of female and male gametes and the determination of gender at the molecular level. It may be surprising to learn that much of our current understanding of molecular biology, the chemical triggers for early embryonic development, many regulatory processes in genetics and body pattern formation, and a host of other fundamental biological processes, have been derived from bacteria, yeasts, roundworms, or fruit fly models.

The first recorded observation of root-knot nematodes dates back to 1855, when the Rev. Miles Joseph Berkeley, using a microscope, found tiny worms in root galls of greenhouse-grown cucumber plants. The scientific name of the organism underwent several changes until it was re-renamed (it's a long, drawn-out story) *Meloidogyne* by nematologist Benjamin Chitwood in 1949. In all species of host plants infected by the many different species of *Meloidogyne*, the characteristic underground "galls" or "knots" on the roots are good indicators of infection. Above-ground, the plant may lack growth, it is prone to premature wilting, will not set or produce fruit, or may even die.

Correct identification of the root swelling is important, because not all galls are caused by nematodes. The determination of the species of *Meloidogyne* is also crucial. That's because one way to deal with a field infested with nematodes is to use crop rotation. But some species of the root-knot nematode have extremely broad host ranges, and planting different crops in succession in the same field may not solve the problem.

One of the high-value crops that can be decimated by *Meloidogyne* is tomato. A lot of work has gone into breeding nematode-resistant tomato varieties. Almost all of the resistance genes bred into the cultivated tomato are derived from a wild ancestor, *Lycopersicon peruvianum*, a plant that produces small, inedible fruit and is naturally found in the Andes Mountains. One of the reasons that it is very useful to know the places of origins of crops, and then catalogue and save seeds from ancestral plants, is that these may possess valuable genetic traits that were accidentally "lost" during the process of breeding for desirable characteristics, like fruit size, taste, or nutrient content. As correctly predicted by Vavilov (please see the essay, "A Different Kind of Bank"), the geographic origins of plants are likely to harbor ancestral plants with diverse and potentially valuable genes.

Genetic engineering is currently the rave of all things related to disease resistance. But nature has a better – though not necessarily faster or easier – solution: a deep reservoir of time-tested genetic material from ancient plants that co-evolved with their enemies. We are constantly told that genetically engineered organisms are safe. Maybe, they are. Then, again, scientists in the 1940s also predicted that all bacterial diseases would be conquered by antibiotics by 1960; they told us that DDT and other pesticides were perfectly safe and only targeted pests; that nuclear energy can be used not only to generate electricity, but that it would be used to "dig" tunnels, widen canals, and blast new building sites; and many other predictions that not only didn't work, but have actually backfired. We, humans, ought to have just a bit of humility. Life has been around for three plus billion years. We have been around for a blink of an eye of that time.

# A LONG WINTER'S SLEEP

*We animals are the most complicated things in the universe.*
*The universe that we know of, of course, is a tiny*
*fragment of the actual universe.*
*-Richard Dawkins*

Every spring I get a woodchuck to take up residence under my barn. I don't mind this; in fact, I like watching him from the kitchen window out for his leisurely morning meal of clover, grass and herbs. Unfortunately, our association typically doesn't last for very long because of the cat. Oh, Mr. Farley (I named him for a character in one of John D. MacDonald's Travis McGee series because of his "fearful yellow eyes") doesn't tangle with the woodchuck. He knows better than to attack one with such formidable incisors, sharp claws and bad temper. But what Mr. Farley's slinking does is to annoy and disturb this wary and rather private creature until it decides to move on to more quiet surroundings. Because of this, I am often forced to watch the woodchuck from a distance throughout the summer months.

What is a rather skinny, drab, and pitiful-looking animal in the spring becomes handsome, shiny and fat by the fall. It's a good thing that the groundhog has gained so much weight because this fat reserve is going to have to carry him through the winter months. That's because the woodchuck is one of the few true hibernating warm-blooded animals.

Ask a hundred people what animal they think of first when you mention the word hibernation, and most will say "bear." However, the North American black bear and the European brown bear are not "true" hibernators. Although bears spend part of the winter sleeping in their dens, they often wake up and occasionally venture out to stretch, feed or urinate. More importantly, the severely reduced metabolic activity of the body, which is the hallmark of true hibernation, is not seen among these mammals. "True" hibernators, like the groundhog, non-migratory bats, jumping mice and some species of western ground squirrels, sleep right through the winter and do not venture outside their homes until the spring.

The difference between true hibernation and periodic torpidity is much more than just the length of the resting period. Partially dormant animals, like the bear, can lower their body temperatures by a few degrees during their rest periods (actually, humans also experience this as well in a much more abbreviated fashion as our body temperature drops by a degree or two during the wee hours of the morning).

True hibernators, however, can drop their body temperatures much, much lower. According to Peter W. Hanney's excellent book, *Rodents – Their Lives and Habits,* "the body temperature of a hibernating ground squirrel is...30° Celsius [about 85° Fahrenheit] below its normal body temperature." Groundhogs, too, maintain a body temperature of about 40° F during their sleep compared with their normal body warmth of about 100° F.

Along with reduced temperature, their heart rate and breathing are lowered by about 95 percent. Interestingly, even the true hibernator may warm its body temperature periodically. Ground squirrels, for example, have been shown to rapidly warm their body temperatures for several hours before returning to a fully torpid state. And they can do this quickly: in one experiment, the animal's body warmed from 38° F to 99° F in 3 hours, only to return to its nearly frozen condition in a few more hours' time. Why or how they do this is yet to be discovered, but scientists are keenly interested in the mechanisms of achieving these altered physical states.

Many of the true hibernators produce two kinds of fats during the summer months: regular fat, which is an energy storage unit, and brown fat, which is mostly used for heat generation during hibernation. The brown fat is burned off very slowly in the winter in order to keep the body temperature from falling below a critical point.

Not only are nutrients used up slowly by this reduced metabolic activity, but build-up of toxic wastes is also curtailed, so that the animal doesn't have to wake up to relieve itself. It also may escape detection by potential predators, many of whom find their prey by scent or by detecting body heat. Even though the body's functions are at their minimum, the brain of the hibernator remains active and constantly monitors the physiologic state of the body and its surroundings.

With the arrival of spring, "the thermostat [of the brain] apparently sets off a chain of events...and the body warms quickly," writes Hanney. Whereas a comfortably hibernating animal may be next to impossible to wake up, its brain can send signals quickly to revive it at the appropriate time. Just exactly what causes these animals to wake up is not yet fully understood. A combination of loss of food reserves, warming soil temperatures, detection by the brain of other environmental conditions (such as a "smoke detector"-like effect of measuring increased carbon dioxide levels in the den) have been proposed as clues for ending slumber.

Other intriguing possibilities exist. Experiments with the golden-mantled ground squirrel of western North America indicate that the bodies of these animals have annual rhythms associated with hibernation. In one set of experiments, captive ground squirrels were kept in the laboratory under constant temperature and light conditions so that they couldn't detect the change of seasons. Even after years of captivity and no environmental changes, the squirrels maintained a routine of 3-4 months of hibernation before resuming their normal activities. In another experiment, blood serum injected from a hibernating animal into one that was fully active quickly induced the recipient to enter into stupor. Thus, an internal biological clock mechanism was proposed for the observed annual rhythms.

Possibly involving a part of the brain that is sometimes referred to as the "third eye," the pineal gland is suspected to provide the cue for these rhythms. This tiny structure produces several chemicals, the most important of which is a blood circulated hormone called melatonin. The same substance, produced by the same structure in the human brain, that has been implicated in seasonal depression in some people, known as seasonal affective disorder, or SAD.

Melatonin production in humans changes with the seasons and also fluctuates daily, with the highest level occurring at night and its ebb is at midday. Although buried deeply within the brain, the pineal gland does receive signals from the eyes regarding light conditions. Interestingly, the melatonin made by the pineal gland has a huge effect on another part of the brain, the hypothalamus. The latter structure maintains many key body functions, including (but not limited to) regulation of food intake, body temperature monitoring and regulation (the "thermostat" effect that Hanney referred to), and control of sleep-wake cycles. It seems that biological clocks are present in many, if not all living creatures and govern their daily, seasonal, and yearly activities.

I doubt whether groundhogs know or care about melatonin, brain parts or such. Of one thing we can be sure: my little friends will be spending the summer getting fat and digging holes and tunnels and enlarging their dens, much to the chagrin of horse owners, farmers and gardeners.

As for me, I will be watching the woodchucks through my binoculars, and Mr. Farley will do his level best and continue to pretend to ignore them. But if they get into my Swiss chard, like one enterprising woodchuck did last year, I am going to have to implore Mr. Farley to take a more aggressive stance or, more likely, rethink my strategy and trap and relocate the little fellow to greener pastures!

# A STONE, A WINDOWPANE AND
# THE NIGHTLY NEWS

Three minor incidents that happened on the same day made me think of…bacteria. Oh, not just any old germ, but a very special kind of microorganism. The events that led me to think about these microbes were: working in the garden; looking at my reflection in a window; and watching a television news story. What in tarnation is the connection among these seemingly unrelated occurrences? The common thread is a group of bacteria called the Clostridia.

OK, let's start at the top and see how my weird mind got me thinking about these critters. First, working in the garden. I was collecting stones, the kind of flat pieces of shale that are in great abundance in my yard every spring. In fact, although I have not had a lot of success growing vegetables until I built raised beds, I seem to have a great crop of these rocks each spring. Frost-heaving, I think it's called. The harsh winters of western New York are perfect for disgorging long-buried rocks of every shape and size from the earth. So, every year before I begin the seasonal task of mowing, I collect the stones unearthed during the winter months and stack them around my flower beds. Performing this annual ritual, I thought of the Wendell Berry poem, "The Stones":

*I owned a slope full of stones.*
*Like buried pianos, they lay in the ground,*
*shards of old sea-ledges, stumbling blocks*
*where the earth caught and kept them*
*dark, an old music mute in them*
*that my head keeps now I have dug them out.*


Distracted, I dropped one of these stones and, as I instinctively tried to grab it, I cut myself. The flap of damaged skin dangled from my knuckle and blood was dripping from the quarter-sized wound. "I best clean the dirt from this cut," I thought, while also trying to remember when I had my last tetanus booster.

The second incident occurred later while I was washing the living room windows. I happened to catch a reflection of myself in the glass and it wasn't a pretty sight. Lines in my forehead now run both vertically and horizontally. From my left eyebrow to the ever-receding hairline, there is now a long furrow extending at an oblique angle. There are bags of puffy skin under my eyes. Too many years in the sun; too many cigarettes smoked; too many bottles of whiskey drunk over the decades. "I sure could use some cosmetic work, maybe like botox to relax the wrinkles away," my vanity exclaimed.

That same evening I saw a news story about botox. According to the report, a married couple wanted to look younger and more attractive and decided on botox injections. Only, the preparation wasn't what it was supposed to be and, instead of gaining a more youthful appearance, both individuals almost died from the fake or contaminated product. After months in the hospital – much of it spent on life support – they were finally able to tell their story to the TV reporter. "So much for botox injections," I thought. These three incidents led me to think about a group of common bacteria, the Clostridia. So, you ask, what is the connection?

Members of the genus *Clostridium* are rod-shaped single cells that stain dark blue with the most commonly used dye for coloring bacteria – this makes them what is known as Gram positive. Another characteristic of this group is that they live without oxygen, which is actually harmful for them. And these bacteria are spore-formers. Under certain conditions the cells are capable of producing an extremely weather-resistant resting structure known as an endospore.

There are about 120 species in the genus *Clostridium*. They are common residents in soils, in sediments and even in the intestines of many kinds of mammals. Most are harmless and some are even beneficial. For example, some species were used at one time for the industrial-scale manufacture of acetone and butanol. Several species can "fix" nitrogen (which, unlike oxygen, doesn't harm them) from air and make it available in the soil as a plant nutrient. One species breaks down cellulose of plant fibers and is an important decomposer of plant debris buried in the ground.

Of the not so benign species, several are harmful to humans. Two of these have achieved particular notoriety: they are the causal agents of tetanus and botulism. Both organisms produce powerful poisons known as exotoxins, and these are the ones that I would like to talk about now.

*Clostridium tetani* (from the Greek word tetanus, meaning "to stretch") is responsible for a potentially fatal illness. The bacterium normally lives in the soil, in pockets of airless (anoxic) space. Humans may become infected when a deep, penetrating wound introduces the cells or spores into the body. The proverbial rusty nail is a good example as a point of entry into the body. (The important point is not that the nail is rusty – that merely indicates that the metal has been in the soil long enough for the bacterium to adhere to its surface). The puncture wound that is likely to be caused by stepping barefoot on a nail is such that when it is pulled out, the wound often seals itself. This, then, can lead to an oxygen-poor area in the injured flesh in which the bacterium can live or the endospores can germinate. The injury made by the sharp, pointed object causes localized cell death, further setting the stage for an anoxic environment. The bacterium multiplies in the body for 4-7 days, but rarely moves from the site of the wound. However, its toxic byproduct can spread in the body via blood.

If the toxin comes in contact with certain types of nerve cells, called motor neurons, it can hitch a ride on these to the spinal cord. It is here that the tetanus toxin causes its greatest damage to the body. In the central nervous system, it affects nerve cells that control muscle contraction and relaxation. When these neurons are paralyzed, the continuous release of a chemical, called acetylcholine, keeps the muscle cells from being able to relax. The muscles are then constantly told to contract, resulting in the uncontrollable twitching of the body. Many muscle groups may be affected, including the facial and jaw muscles. This has given the condition its common name of "lockjaw." Unless medical treatment is quickly started, the fatality rate (usually from failure of muscles that control breathing) may reach as high as 70 percent of afflicted individuals.

Fortunately, very few cases of tetanus are reported annually in developed countries because childhood vaccination has become routine. The numbers of tetanus cases occurring world-wide is not well documented but is likely much higher in countries where routine vaccination is not practiced. Even in those individuals who survive the initial attack, recovery is a slow and painful process. If there is any good news, it is that a fully recovered person rarely suffers permanent damage to their muscles or nervous system.

An even more remarkable toxin is produced by *Clostridium botulinum* (from the Greek word "botulis," meaning sausage, as botulism once was thought to be caused by rancid meat). The poison made by this organism is quite possibly the most toxic substance known in nature. One book describes the potency of this protein as "a single microgram of it is a lethal dose for 200,000 mice," and another publication says that "a gram [about 1/30 of an ounce] can kill one million guinea pigs." I don't know about you, but I have trouble visualizing hundreds of thousands or millions of small dead animals. Perhaps, a better example was given by a physician who is quoted as stating that "half a teaspoon could kill all of [the people of] Great Britain."

It would be more correct to refer to the toxin in the plural form, since there are seven different forms of the protein, made by different varieties of the bacterium under different conditions. Known as types A through G, at least four of these (A, B, E and F) can cause illness and or death in humans.

*Clostridium botulinum* is a relative of the tetanus bacterium, but its toxins work very differently than the one that causes "lockjaw." The effect of the botulin is to prohibit muscle contraction. The toxins do this by attaching to nerve cells and preventing the release of acetylcholine. This causes paralysis, since the muscle cells only contract when they are "told" to do so by signals from the nervous system.

So, although both the tetanus and botulism toxins act on the neurotransmitter acetylcholine, the outcome is exactly the opposite: in the case of tetanus, muscles keep contracting, while in the case of botulism, muscles are incapable of doing so. If the muscles that control breathing are affected, an individual may suffocate as a result.

In many other ways, the two bacteria are similar. Both require oxygen-free living space, both form spores, and both produce toxins that can cause death of people and animals. Neither bacterium grows well in the adult human body, although accidentally ingested *Clostridium botulinum* spores can grow in the intestines of very young children, causing what is known as infant botulism. Neither tetanus nor botulism toxin can be spread from person to person.

Botulism is typically the result of eating poorly cooked or preserved food substances. At one time it was only thought to be caused by undercooked or smoked meat products, but it is actually more frequently encountered in improperly preserved vegetables, such as green beans, potatoes, beets, chili peppers, etc. Both home-canned and commercially preserved foods have been implicated in outbreaks of poisoning.

The actual annual numbers of cases of poisoning and death in the U.S. are quite low (typically less than 100 cases of illness, with 5-7 percent mortality). That is in large measure due to several factors. Botulism is one of the so-called "reportable" diseases, meaning that doctors and public health agencies keep tabs on the numbers of cases and report these to the Centers for Disease Control and Prevention. That agency, in turn, educates doctors to correctly diagnose the problem quickly and to take action to prevent death. Other governmental bodies, such as each state's Cooperative Extension Service, also campaign aggressively to inform the public about proper steps of home canning. Finally, the commercial food industry in the U.S. and other developed countries safeguards against foodborne illnesses.

There are no good world-wide estimates as to the numbers of cases of illness or death, but I suspect that there are a lot of unreported cases, especially in countries where most of the population still depends on subsistence agriculture. The good news is that boiling food for about 10 minutes inactivates the toxin. The bad news is that fuel for such thorough boiling is also scarce or very expensive in the poorest countries of the world.

Although the botulin toxin is an extremely potent poison, it has found use in medical treatment. In very diluted doses, it has been used to treat several muscle disorders, such as uncontrollable blinking (blepharospasm) and involuntary twitching of the muscles of the neck, face or jaw (dystonia). In the late 1980s, some plastic surgeons began to use type A botulin toxin to treat wrinkles of the forehead and face. The Food and Drug Administration initially frowned (pun intended!) on this treatment, but by 2002 the agency approved its use to relax furrows and lines in the forehead.

Known as "Botox," the drug's use has become one of the most popular forms of temporary treatment of facial lines and wrinkles. The American Society for Aesthetic Plastic Surgeons reported that about 1.6 million people received botox injections in 2001 at the cost of about $300-500 per treatment. Prices in 2010 ranged from $375 up to $1,000 or more. When handled by experienced doctors, botox reduces age lines by relaxing the muscles that cause the skin to wrinkle. The effect is temporary and injections have to be repeated at 3-4 month intervals.

In some American cities, there are now "botox parties" where patients gather to be treated by a physician. This causes some consternation for those doctors who feel that this approach trivializes the fact that diluted or not, botox remains a powerful chemical that has to be handled properly. As with everything else that has profit-making capability, tainted, impure or even fake products sometimes make it to the market, leading to the kind of nightmare story that I mentioned earlier.

In the case of the cut in my skin, I did not really consider tetanus seriously. This was not the kind of wound in which the bacterium can thrive. At any rate, I had an up-to-date tetanus vaccination. As for botox, any thoughts of vanity evaporated quickly – not so much because of the horror story that I saw on the news, but because my pocketbook could never afford this kind of expense. Along with my thinning hair and gray-white beard, I guess I will just have to accept aging as gracefully as I can.

And, yet…

# A HIDDEN WORLD IN PLAIN VIEW

*Biological diversity is the key to the maintenance*
*of the world as we know it.*
*-Edward O. Wilson*

A few summers ago I visited a friend in Rochester, New York. I parked my car in front of his house and we spent several hours talking and watching a baseball game on television. When I left, I saw that a sticky film had accumulated all over my car. I had parked under a linden (American basswood) tree and the gooey "honeydew" showering down was courtesy of the millions of aphids which gorge themselves with the sugary sap produced by the leaves.

Aphids are tiny insects that feed constantly but cannot use or store all that food in their minuscule bodies. The excess liquid is exuded from them in the form of the honeydew which then drips onto objects below. A close examination of the underside of a leaf of a heavily infested tree will yield hundreds or even thousands of these soft-bodied insects. According to a 1984 book, *Mysteries and Marvels of Insect Life*, a "hectare [about 2.5 acres] of vegetation may support 5 billion aphids which saturate the soil with 2 tons of sugar in the form of honeydew every day." It is no wonder, then, that a freshly washed car may be coated completely in a short time, much to the chagrin of its owner.

86

Entomologists say that the honeydew itself does not damage the paint and washes off easily. However, vinyl or convertible tops have been known to be discolored by "sooty molds" that use the sugary sap as a food source. This situation is somewhat analogous to the moldy discoloration of tile grout in a shower stall.

Thinking about (and cursing) aphids as I pulled into a car wash station reminded me about the incredible diversity of life forms and their complex interactions on leaves. Far from being a sterile surface, each leaf can support a wide variety of large and small creatures, all of which get their food in one way or another from the plant's ability to convert simple chemicals into sugars using the sun's energy (please see the essay, "Nature's Alchemists," on photosynthesis).

Large numbers of harmless bacteria and yeasts live on a thin film of moisture containing sugars and other nutrients produced by the leaf. These bacteria are part of the natural flora of the plant surface and, in many cases, provide a natural form of protection against would-be pathogens. They do this by outcompeting harmful species of bacteria and fungi for space and food, and by producing antibiotics that inhibit the growth of invaders. Pesticides may, in fact, cause more harm than help in the long run by killing the normal microflora of the leaf surface.

The kinds of microbes found on the leaf change as the growing season progresses. As leaves age, the bacteria and fungi that are springtime pioneers yield to other types of colonists later in the season.

A number of insects feast on plant sap and on the resident bacteria and fungi without hurting the foliage. For example, several kinds of beetles eat molds, pollen, bacteria, and nectar found on leaf surfaces. Fungus beetles snack on molds and pollen, and flower beetle larvae live mostly on dead or dying parts of leaves, thus providing housekeeping chores.

Many insects are, of course, injurious to leaves. Adult insects or their immature forms may bite, chew, suck or tunnel into leaves, thereby destroying leaf surfaces and robbing the plant of nutrients. Some insects, however, are useful predators and devour large numbers of parasitic bugs. The harmful aphids, for example, provide food for a wide variety of other insects. They are a favorite food of both the adult and larval forms of the Lady beetle, each of whom may eat as many as 100 or more aphids per day. The beneficial activity of this beetle was recognized a long time ago and, according to some anecdotes, during the Middle Ages they were called "Cows of the Lord" and "Beetles of the Blessed Lady" (hence the name Lady Beetle).

Another aphid killer is the aphis lion. These fierce-looking larvae of the green lacewing decimate aphid colonies. The aphis lion resembles a dragon and uses its long, sickle-shaped curved jaws to hold a helpless aphid while tearing chunks of the soft body of its victim with its mouthparts. The maggots of some flies also eat aphids. A single hoverfly larva, for example, may eat up to 1,000 aphids during its growth period.

Soft-bodied and defenseless against predators, aphids rely on their large numbers for survival. However, they are not entirely without protection. Some species of aphids seem to have formed an alliance with ants in a kind of mutualistic relationship. The ants provide guard duty for the aphids and also collect aphid eggs in the fall and carry them to protected shelters in their nests. In the spring, the ants carry the eggs to young plants or tender plant parts and watch over them until they hatch. In return for this service, the ants are seen to "milk" the aphids for honeydew, which is one of the ants' favorite foods. Both ant and aphid derive some benefit from their association without harm to either.

Like Horton, the elephant, who stumbles upon the entirely new and unseen world of Whoville in the Dr. Seuss children's classic, there is a hidden world to be discovered in unexpected places. Even in the relatively small ecosystem of a leaf surface, a complex relationship, consisting of a world of "eat or be eaten" has been established.

So, the next time you look at a leaf, think of the immense struggle for survival that is happening right before our eyes. If only we could see it...

# THE PRIVATE LIVES OF FERNS

*Sexual reproduction serves, at least in part, to*
*promote variation between parent and offspring.*
*-James L. Gould*

I have several friends and relatives who are physicians. When people find out that they are doctors, requests for medical advice are usually not far away. I kind of know how the doctors feel.

I am not a physician. But my doctorate is in the field of plant pathology and, thus, I deal with diseases of plants. When people learn this, I get questions about house plants, landscape ornamentals, forest trees, and even inquiries about garden pests. Over the years I have dealt with everything from lace bugs on azaleas to heart rot of conifers, bronze birch borers, gypsy moths, powdery mildew of roses, and "artillery fungi," to list just a few examples.

One of the most frequently asked questions goes something like this: "I have a nice fern in my living room but I think it has an insect problem. There are these blackish-looking bugs on the back sides of the leaves. What can I spray to get rid of these pests?" I usually ask if these "bugs" are all in orderly rows on the underside of the leaf, but I already know what the answer is likely to be before I ask the

90

question. These evenly spaced and neatly arranged dark bodies are not insects but packets of spores. For the fern they represent a means of dispersal in nature because these spores are scattered by wind and, if they land in suitable habitats, they produce tiny plants on which reproductive units will eventually develop.

Most people learned about these structures in high school or college biology classes during a usually brief survey of the plant kingdom. However, the teacher's emphasis typically was on requiring memorization of the life cycle, which included strange tongue-twister words, like "sporangia," "archegonia," "prothallus," and "gametophyte." As a consequence, students may have reluctantly memorized these terms just long enough to take a test, and then promptly forgot about them. And who can blame the student? The focus was not on understanding or appreciating the beauty and connectedness of living organisms, but on requiring the student to regurgitate nearly meaningless words. As a college biology instructor I have been guilty of doing just that. Late in my teaching career I realized the fallacy of this approach and tried to purge my lectures of the superfluous material as much as possible, especially for non-biology majors.

Ferns today are represented by about 11,000 to 12,000 species world-wide. Other than house plants or ornamental plantings they are not economically or ecologically important members of the plant kingdom. Compare their numbers with the nearly 300,000 known species of flowering plants, some of which also include the world's most important food crops (please see the essay, "A Different Kind of Bank"), and the paucity of fern species becomes even more pronounced.

But if you used electricity today, then you need to thank ferns and other fern "allies." That's because in most countries of the world much of the electricity is generated by burning coal to heat water to steam that turns turbines that produce the electric current. And most of the coal deposits mined today were formed millions of years ago by the partial decomposition of giant fern, "horsetail," club moss, and scale tree forests. The fossilized remnants of these plants that lived on Earth during the so-called Carboniferous or Pennsylvanian Period, have lain largely dormant in coal seams for more than 300 million years until recently.

Paleontologists study extinct organisms that lived a very long time ago. Based on geologic evidence from rocks and fossils, it has been postulated that the climates of China, North Africa, Europe, and North America were tropical to sub-tropical and huge areas were covered by swamps during the Carboniferous Period. It was in the extensive swamp forests that the aforementioned plants lived and died, and underwent the slow process of coal and methane gas formation.

As fascinating and important as the processes of formation of energy-rich compounds from once-living organisms are, I am more interested in the fact that the methods of sexual reproduction are very similar among plants, fungi, and animals. By this, I do not mean the steps of copulation, because that is simply a means of bringing male and female units together, and the methods of achieving that can vary widely among different organisms. Rather, I am referring to the fact that even among the "simplest" plants and animals, sexual reproduction involves the formation of (an) egg(s) which is (are) then fertilized by a sperm cell (it is also inconsequential that when discussing different organisms, the names applied to these units may be different).

In ferns, male and female reproductive units (gametes) are formed on a tiny, heart-shaped plant that is nearly impossible to see without a microscope. This small plant arises from the spore that is sometimes mistaken for bugs. Many, though not all ferns, will self-fertilize. The process involves a swimming sperm (hence, at least some moisture is necessary) moving toward a stationary egg. The male gamete is just like the sperm of an animal, including the human animal. The sperm is simply a "delivery device." In order to be able to swim, the sperm is equipped with a whip-like structure, the flagellum, which acts both as propeller and rudder to guide it to its destination (please see the Foreword section for a brief mention of this structure in connection with the 2016 Nobel Prize in chemistry.)

I mentioned elsewhere in this book that one of the basic principles of biology is that very complicated processes are usually solved once in evolution and, if the solution is effective, it is used by a wide variety of organisms. This is the case with bringing two gametes together in sexual reproduction. A large egg, with half of the chromosomes, is fertilized by a tiny motile sperm cell bringing the other set of chromosomes to form the embryo. The fertilized egg then undergoes development and ultimately "grows up" to be the mature plant that we see in the fields, woods and bogs, where ferns typically grow. The life cycle then starts all over again with the production of spores on the underside of the leaf.

Interestingly, biologists still argue today as to *why* sexual reproduction evolved. Not all organisms reproduce sexually (bacteria never do) and even among some other creatures sexual reproduction may be a rare phenomenon (please see the story, "Roundworms Galore.") However, it is clear that sex must have some major advantages. Otherwise, the vast majority of living organisms wouldn't bother with it.

The life of a fern

# FLY, FLY AWAY

*Toward the approach of the day...the Pigeons began
to move off in a direction quite different from that
in which they had arrived the evening before, and
at sunrise all that were able to fly had disappeared.*
-John James Audubon

Here in western New York I always feel a little bit sad around the third week of August. No, it's not because I have to go back to the classroom after a summer of mostly bohemian existence. By this time, I'm ready to meet the new crop of eager (?) students and get back to work.

My melancholy stems from this: I wake up one morning and there is a deafening silence. Instead of the cacophony of bird sounds that I became accustomed to listening to over the summer months, now there is quiet. Gone are the nervous and edgy barn swallows, the "pugnacious" (Roger Tory Peterson's word but very true) hummingbirds, and my personal favorites around the yard – the redwing blackbirds. Soon, the robins will be gone. Another month and the Canada geese will be wildly honking overhead as they fly south for the winter. Then the cat, too, can scratch from his long to-do list some of the birds that he was so fond of stalking.

As I begin the annual rituals of washing and putting away the hummingbird feeders, cleaning out the seed bins and restocking them for the winter, I always wonder how these travelers find their ways to their winter destinations. Poets and naturalist, casting their eyes to the sky, have for eons asked the same questions about the homing abilities of birds. Ethologists, people who study animal behavior, have begun to sort out the answers to this mystery.

Migratory behavior is common among birds and many other animals. Nearly two-thirds of the bird species residing and breeding in North America during the summer months travel south for the winter. Some kinds of birds move only relatively short distances from their nesting sites, while others are transcontinental migrants. The long-distance champion has to be the Arctic tern, which flies more than 25,000 miles from the Canadian North country all the way to the vicinity of the Antarctic Pole. In the spring, it returns to its breeding grounds near the Arctic Circle.

Although such a distance is mindboggling, I am even more impressed by the tiny ruby-throated hummingbirds. The ones that crowd around my feeders in the summer, and are brazen enough that they could care less if I am sitting within 2-3 feet of them watching their every move, fly off to Mexico or somewhere in Central America at the end of the season. How can such a tiny creature, weighing a mere tenth of an ounce (3 grams), make such a journey without perishing along the way or getting completely lost? Equally importantly, how do they find their way back to my house next summer?

Surely, they must use some sort of navigation system accurately because, over such long distances, even small errors in the direction of flight may put a bird several hundreds of miles away from its destination. A number of hypotheses – some backed up by experiments – have been proposed to explain the navigational capabilities of birds and other animals.

One of the first individuals to systematically study the homing abilities of birds was the British ornithologist, R. M. Lockley. He and his co-workers captured adult Manx shearwaters (a kind of seagull) at their nesting sites off the west coast of England and transported them to various sites before setting them free. The birds were kept in closed boxes until their release from rooftops in Cambridge and Birmingham in the U.K., from the Swiss Alps, from Venice, Italy, from a ship in the Faroe Islands, and from the Boston airport in the U.S. Most of the captives were able to quickly orient themselves in the correct direction and return to their nesting sites within 2-12 days (depending on the distance transported) if the sun was out or if there was at least some clear sky. However, if the weather was completely overcast the birds appeared to be disoriented and some would not even attempt to start on their return flight until the sky cleared.

So, it seemed that the birds needed certain clues to position themselves in the correct direction. The most obvious cue appeared to be the sun. The first person to provide experimental evidence for the sun's importance as a navigational aid was Gustav Kramer in the 1950s. He observed that European starlings exhibited migratory behavior when placed outdoors in circular cages during the times of the year when these birds normally take off on their journeys. Their activities were oriented in the correct migratory direction in both the spring and the fall as long as the sun was visible to them. Orientation was poor if the sky was completely overcast. Furthermore, the directional choices of the starlings were altered when the true position of the sun was deflected by mirrors.

But to use the sun as a compass during migration, the birds must account for the movement of the sun across the sky. In other words, the animal must change its orientation to correctly compensate for the earth's rotation. Tests with pigeons, starlings and owls have shown that they are equipped with light-sensitive eye membranes and are able to make precise measurements of the sun's position in the sky and change their flight orientation depending on the time of the day. These birds use the sun not only as a compass, but also as a sun dial to determine the time of day, and even as a sextant to measure latitude.

Not all birds are daytime travelers and some do not stop to rest when the sun goes down during their migration. Therefore, orientation by means other than the sun must exist among some species. Night time fliers can and do make use of star maps and this has been demonstrated by many clock-shift experiments, and by varying the stellar maps that birds can see.

In one experiment, indigo buntings, red-backed shrikes and Blackcaps, time-shifted by 3-12 hours under artificial light regimes were exposed to planetarium skies. If the birds were using internal clocks to read the sky, time-shifted birds should have made significant errors in flight direction directly related to the number of hours of the shift (i.e. 15 degrees error per hour). This was not the case, however, and there was no significant deviation from the correct migratory direction. Therefore, the birds were "reading" the star map directly.

Planetarium experiments by Franz Sauer further showed that birds use the location of stars for orientation. Warblers shown star patterns coinciding with autumn migration became restless and took up positions in their cages in the correct flight position. The cages then were covered while the whole planetarium sky was turned 90 degrees counterclockwise, so that the stars which should have been pointing to the south were now pointing east. The cages were then uncovered and, within a few seconds, the birds re-oriented themselves in the new false north-south direction. Many nocturnal migrants refuse to take off under total cloud cover. Even those that do, exhibit reduced accuracy in flight direction. Birds already flying get confused and lost if they can't see the stars.

There is also some evidence that many animals, including birds, are able to detect tiny differences in the earth's magnetic field. (This is true even for bacteria: please see the story, "I'm Drawn To You," elsewhere in this book). There appear to be specialized magnetic sensors in the brains of pigeons and other birds, allowing them to determine the angle at which magnetic fields are created in the earth's fluid core and using this information for orientation purposes. These data must then be stored in the brain for future use.

Memory is also involved in those species that use various landmarks as visual cues for travel. (We are, then, talking about storing information in the brain which must be accurately remembered and recalled. So much for the term "bird brain" as a put-down!). Canada geese and homing pigeons routinely use topographic features, such as mountains, lakes, rivers, buildings, or smokestacks for short-distance orientation. However, such landmarks are not absolutely necessary for these birds to find their way, and may be used by them only when they are conveniently available.

There are other tantalizing lines of evidence that indicate that directional clues are derived from taste and smell, which are highly developed in many animals; from detection of changes in barometric pressure; from presence of sonar sensors; and by measurements of the earth's rotational force.

All of the ideas concerning the navigational abilities of birds attempt to explain the concept through a coordinate mapping system. This, of course, is how humans would approach the problem of direction finding.

Most likely, birds use a combination of different methods during their annual journeys. We are just now starting to explore and understand a process that birds, butterflies, sharks, whales and many types of insects have been using for millions of years. They have had global positioning systems (GPS) all along! And now we, humans, come along in the 21st century and congratulate ourselves about how clever we are with all the new satellite-based technology!

All I know is that come next April, my many feathered friends are going to be back. Then, I will begin to put out the feeders and begin making the sweet sugary syrup for the hummingbirds. The cat will once again look longingly at the feeders and dream of catching an unlucky or careless victim.

# SPRING SYMPHONY

*The odor of a ploughed field in the spring is
like a hand laid upon the heart...*
*-Henry Beston*

No, this is not a story about Robert Schumann's or Benjamin Britten's symphonies, or any of the other musical compositions by that title. It is a story about frogs.

Spring comes late to western New York. During early March I see brave green-purple skunk cabbage tips peeking through the snow in swales in the woods, but these are just as quickly buried by a fresh layer of white. In mid-March, I usually spy the first robin, a rather disoriented and disheveled-looking male, in the back yard. It doesn't have much to eat, yet, as insects rarely venture out into the cold wind that mocks the spring equinox. Sugar maples set their tiny and inconspicuous flowers, and the red maples are not far behind. Mercifully, the days are getting longer.

But the true harbingers of spring are the peepers. Well after the sun begins to descend in the western sky, the first few spring peepers tentatively begin their chorus. They sound unsure of themselves after a long winter's slumber. But if the weather stays decent, their voices grow louder, more confident, until a few days later the cacophony becomes deafening.

Spring peepers are tiny tree frogs. The males are generally about an inch (2.5 cm) long; the females are slightly larger. They spend the cold winter months buried under leaf litter and moss or, protected by snow, in and under logs and tree roots. Since they are "cold-blooded" (i.e. they cannot regulate their body temperature), they are at the mercy of their surroundings and, in some years, may not emerge until early to mid-April.

Most of us never see these tiny, secretive nighttime creatures, but those living near marshy areas or vernal pools can't help hearing them. The sometimes overwhelming chorus of hundreds of males, vying for a chance to mate with choosy females, begins with the waning hours of daylight and continues well into the night. Vocalization is by males only, and includes the whistling peep of their amorous advances, and a throaty, unmusical call when competing with other males for territorial space.

Peepers are not the only frogs mating in the spring. Wood frogs, green frogs, leopard frogs, and several species of toads also get on with the business of making babies. Because different species of frogs prefer different bodies of water for their reproduction, the mating habits and seasons of different frogs, although overlapping, are different.

Water is critical for reproduction for these amphibians. Even though males copulate with females, eggs and sperm are shed into water, where fertilization and development of tadpoles take place.

Many species of frogs, toads and other amphibians have declined in health, population size, and distribution throughout the world as a result of water pollution and habitat destruction.

J.A. Allen wrote this comment on the loss of reptiles and amphibians of the Springfield, Massachusetts, area: "Draining ponds and marshes, destruction of the forests, and burning of newly-cleared lands [has reduced]...many species to great rarities." His notes were penned in 1868.

Fortunately, judging from the roar of their joyous harmonizing in the spring, the spring peeper is alive and well in 2016. That is due, in large part, to nature reclaiming former agricultural and logged-over lands. New York State, for example has gone from less than 25 percent forested land in the early 1900s, to more than 60 percent forests in the 21st century.

The story, "Rebirth," elsewhere in this book chronicles the restoration of wetlands. The Clean Water Act of the 1970s has been instrumental in cleaning up ailing lakes, rivers and streams and the restoration of diverse wildlife habitats. Hopefully, these efforts will continue well into the future, as we need to heal ecosystems that we have so successfully destroyed during the last few hundred years.

# AUTUMN DELIGHTS

*Mushrooms can be every bit as beautiful as birds,*
*butterflies, shells, and flowers, yet we never think*
*to describe them in such flattering terms.*
*-David Arora*

There is no question about it – autumn is the best time of year. I don't like the sweltering and sweaty heat of July and August, and the short, cloudy days of winter often depress me. Autumn, on the other hand, is wonderful. Splendid leaf colors of crimson red and ochre yellow, burnt umber and pale shades of green are framed by a cerulean September sky.

I never fail to marvel at beauty of the sumacs whose leaves take on a thousand hues of green, red and purple. After the first frost the pesky mosquitoes and those tiny gnats that always seem to end up in my eyes disappear. The skies are crowded with the loud honking of Canada geese. Huge flocks of noisy blackbirds invade my back yard at early dusk to roost for the night. The nights are cold enough to burn a few logs in the wood stove but the days are still pleasantly warm and comfortable.

Another great thing about the season is the abundance of mushrooms. They are everywhere – on the lawn, in the woods on the forest floor, on logs and standing trees, in the garden on rotting straw and piles of grass clippings. They come in all shapes and sizes and colors. I don't pick and eat the mushrooms because I don't consider myself expert enough to know the "good" ones from the "bad." But I do enjoy looking at the immense variety of fungi living within a stone's throw of my front door.

Each autumn day brings a new excitement of discovery. Where there were no mushrooms the day before, suddenly a big clump of them appears the next morning under the spruce and aspens, or colorful colonies miraculously spring up overnight on a decaying log. Sometimes, circular patches of "fairy rings" emerge on the lawn but then they are gone just as quickly. After a good, soaking rain and cool nights the woods are teeming with a great variety of toadstools and tiny, colorful slime molds. But their presence is usually ephemeral and the mushroom hunter had better hurry to find them. The only ones with longevity are the hard, leathery or woody bodies of conks and other shelf fungi growing on the trunks of living trees or on crumbling woody stumps.

The seemingly sudden appearance of mushrooms out of the ground has given rise to many folk tales and myths. In many cultures throughout the centuries the origins of fungi had been attributed to supernatural forces. Ancient Indian, Greek, and Roman cultures sometimes elevated fungi to godly status and worship rituals were not uncommon. For example, the Romans offered animal sacrifices during the annual celebration of Robigalia to appease the rust god, Robigo. "Rust," a fungal disease of cereal crops is a serious problem even today. For the Romans, a bad season of rust outbreak meant starvation and Robigo held the key to a harvest of plenty or a year of famine.

Fungi also have been used to support the argument for "spontaneous generation," a concept that life can appear out of nothing. This view was still accepted as a valid scientific principle until the latter part of the 19th century. It took the likes of Anton DeBary, Robert Koch and Louis Pasteur to finally put that myth to rest.

Actually, the above-ground or visible part constitutes only a small portion of the body of the fungus. For most of the year, the mushroom lives underground, existing and growing by small, thread-like strands. These filaments, called hyphae, make up what is collectively known as "mycelium." The mycelium (from the Greek word "mykos", or "cap", where the term for the study of fungi, known as mycology, originated) may make up hundreds of miles worth of intertwined living tissues. Under the right environmental conditions, the mycelium begins to produce the mushroom.

Although fungi lack true roots and stems, a stalk and cap rapidly emerge from the soil or growing medium. In the gills, pores or labyrinths of the cap, trillions of spores are produced which are then dispersed by wind or water, or they may stick to the bodies of animals who carry the spores to new environments. If they don't land in a suitable location, the spores of some fungi may lie dormant for a long time before germinating to start a new colony. Usually, two different but sexually compatible strands of the fungus have to grow together before the mushroom caps and spores are produced.

Therefore, there is "true" sexual reproduction in these organisms. Many fungi, however, also can reproduce by simple cell division or by breaking up the mycelium into smaller fragments.

Next to insects, fungi are probably the most numerous and diverse group of organisms. Scientists estimate that there are over 1.5 million species, of which only about 10 percent are known. Most of us are familiar with only a few dozen or so types of fungi – those green or black molds that ruin the loaf of bread on top of the refrigerator; the mildews that grow on the shower curtain and bathroom tile grout; the mushrooms we buy in the store; and the occasional "toadstools" or "conks" we may encounter on a nature walk. But fungi come in all shapes and sizes and colors and textures and they occur just about everywhere. Many are quite unique and do not in any way resemble the common mushroom we are familiar with.

One of the most interesting types is the group of "bird's nest" fungi which, as their name implies, resemble miniature nests complete with "eggs." The egg actually contains spores which are dispersed when the force of raindrops "cracks" the "shell" and the spores are forcibly discharged from the nest-like cup.

Another fungus that shoots out spores is aptly named the "artillery fungus." This critter is often found growing in mulch, which may be used to cover the soil of foundation plants near the side of a building. The fungus is able to propel its black spores for up to 5-6 meters (15-20 feet) which stick to siding, porch furniture, or cars, much to the horror or chagrin of the home owner. I have had several calls over the years from people who wondered what all those black specks were and where they came from.

Yet another interesting group of fungi is the puffballs. Their spherical bodies lack a stalk and can grow quite large, occasionally reaching up to 30 inches in diameter and weighing nearly 35 lbs.

Some mushrooms are legendary for their flavor and are sought after as great delicacies. Truffles, for example, fall into this category and exclusive restaurants pay handsomely for this elusive fungus. Since they grow underground, truffle hunters use specially trained dogs or pigs to sniff them out. Pigs absolutely love truffles and unless their noses are pierced and outfitted with a large ring, they will dig up and quickly devour them. A truffle hunter will guard the location of his prize like a trade secret.

Other mushrooms are good to eat when they are young but become bitter or undigestible as they age. Many of the puffballs are edible as long as their inside is white; when the spore masses turn brown it is too late to eat them. The bright yellow shelf fungus, *Laetiporus sulphureus*, is sometimes called the "chicken of the woods" for a good reason. It is quite delicious when it first appears on a fallen log and it is often sautéed in butter or oil with garlic or chives. It becomes tough and inedible after a few weeks' growth.

Certain types of fungi produce deadly toxins. Some species of *Amanita* are known as "death caps" or "destroying angels" due to their poisons. Contrary to folklore, silver spoons don't turn black when a mushroom is poisonous. It's best to be familiar with the ones that kill!

Still other fungi produce hallucinogenic compounds, like psilocin, psilocybin and LSD. I once sat with a friend for an entire afternoon to reassure him every few minutes that alien forces weren't about to kidnap him; he had bought and ate some "shrooms" at the rock concert we attended.

While fungi are generally not held in high regard, they are very much part of most ecosystems. In the forest, they are decomposers of fallen leaves, cast-off branches, insect-weakened or dead trees and uprooted logs, and they recycle nutrients back into the soil.

Many fungi are found only under certain specific trees. These mycorrhizal fungi have a mutualistic association with their host tree species. In a you-scratch-my-back-I'll-scratch-yours relationship, the fungal mycelium greatly enlarges the roots' surface area for absorption of water and minerals from the soil; in return, the tree supplies nutrients for the fungus.

I often think of these things as I'm hiking in the woods on those balmy autumn days free of bugs. I savor the earthy smell and the sensations of color and texture of the different fungi I encounter in the forest. But, then again, I already confessed elsewhere in this book that I am a science nerd.

# I'M DRAWN TO YOU

*Nothing is worldwide, but everything is episodic. In other words, the history of any one part of the Earth, like the life of a soldier, consists of long periods of boredom and short periods of terror.*
*-Derek Ager*

North is north, and south is south. End of discussion, right? Well…yes and no. Planet Earth clearly has geographic north and south poles. That being the case, Earth also has an equator – by convention 0 degrees latitude, whereas the poles are 90 degrees north and south, respectively. But the planet also has a magnetic north and south, as anyone with a compass can attest to. At the present time, the magnetic and geographic poles more-or-less coincide but this was not always the case. Even today, the magnetic poles "wander" and their locations have to be reconfigured every few decades. How can this be?

The currently most accepted model of our planet is that there is a solid inner core of mostly iron-nickel composition, along with some sulfur, at the depth of about 5,000-6,000 km (3,000-3,600 miles) below the surface. The solid core is surrounded by a liquid outer core of a sea of molten metal. Outside of that are the solid mantle (made up of magnesium and iron) and the crust, the outermost layer, upon which we stand, of complex mineral composition. At least, that's what geologists think.

Since it's impossible to see to the center, the major clues for the Earth's interior were initially provided by meteorites. The 19th century American geologist, James Dana, looked at the chemical make-up of different kinds of meteorites and postulated that the Earth's interior may be composed of similar kinds of metallic or rock materials. More recently, rocks from the Moon, and spacecraft data from Venus and Mars seem to confirm Dana's conclusions. Further evidence for the solid-liquid-solid arrangement comes from seismic information and measurements of the Earth's magnetic field.

That there is a magnetic field associated with the planet has been recognized for hundreds of years when William Gilbert concluded in 1600 A.D. that the Earth is magnetic. The cause of magnetism was first suggested in 1820 by Andre-Marie Ampere, the French physicist who proposed the presence of internal electric currents deep underground.

But what causes these electric currents in the first place? In the 1930s, physicist Walter Elasser suggested that the currents were generated in the liquid outer core, which is an electrically conducting fluid in constant motion. According to Elasser's hypothesis, the motion is due to heat convection from the molten metals, much like a pot of boiling water creates bubbles, swirls and eddies in a complex roiling motion. According to Elasser's "dynamo hypothesis," the magnetic field is the result of interaction between the differential rotation of the solid mantle (faster) and the slower fluid core outside of a more rapidly moving inner core. The Russian scientist, Stanislaw Braginsky, offered mathematical proof in 1964 for the possibility of Elasser's proposal. More recently, the Earth's spin about its own axis, creating the Coriolis Effect, and the fact that there is a "wobble" in the spin (called precession) also have been implicated in creating magnetism.

Okay, so far, so good. So, there is a geographic north and south and a corresponding magnetic north and south. Not so fast. The fly in the ointment is this: the strength of the magnetic field waxes and wanes, and the poles have not only drifted but apparently actually reversed themselves many times in the long geologic history of the planet.

Evidence for the polarity reversals has been preserved in rock layers 400-500 million years old and there is no reason to suspect that such phenomena did not occur long before that. The best documented history, however, is found in much younger rocks of the late Cenozoic Era, going back to 4-5 million years before the present. Data show that lava rocks laid down during volcanic eruptions, or deep-sea sedimentary layers that precipitated out of water, contain minerals that have preserved evidence of many previous magnetic reversals.

These events seem to occur randomly. During the Cenozoic, the reversals have been as long as a million years apart and as short as 10,000-year intervals. Most flips are estimated to take a few thousand years to complete. The strength of the magnetic field – the so-called "dipole moment" – varies considerably during reversals, but it does not disappear altogether. The most recent complete magnetic flip occurred about 800,000 years ago. The less dramatic drifting of the magnetic poles continues, however, currently at about 10-40 km per year.

It stands to reason that the inhabitants of the planet are affected by the Earth's physical characteristics, including magnetic forces. As mentioned in the essay on bird migration, titled "Fly, Fly Away," some animals use information about magnetic forces to orient themselves during long-distance travel. Do these animals, then, get confused by shifting magnetic north-south directions? Perhaps not. The magnetic shifts are spread out over a number of years and may not occur significantly within a single animal's lifetime. Even some long-lived species, like sea turtles, gather information from a variety of cues to correct for any error. Some scientists have speculated that the well-documented extinctions during geologic times may have occurred during magnetic reversals. However, there is no solid evidence to back up such hypotheses.

On the other hand, there are creatures that are profoundly and rapidly affected by magnetic fields. I first read about these while I was a master's degree program student at West Chester State College (now University) in Pennsylvania in the early 1980s and I was completely fascinated by them. Apparently, there are lots of different aquatic bacteria that have tiny magnets in their cells and they can use these to orient themselves along magnetic fields. They are called magnetotactic bacteria (taxis = movement). The actual motility is provided by tiny, whip-like structures called flagella that act as "outboard motors" to propel the cell (please see my comments in the foreword about molecular motors and the 2016 Nobel Prize in Chemistry).

How can these bacteria know what to do? That was the question asked by Richard Blakemore when he first isolated them from a sample of pond water in the 1970s. And, of course, the "why" question soon followed: what purpose would be served by the ability to orient along magnetic lines?

As for the ability to act as a compass, all magnetotactic bacteria contain either magnetite (an iron oxide) or an iron-sulfur substance, called greigite. There is even a species that has both minerals. These microscopic bars of magnetic material are packaged into tiny pouches, making up what are referred to as magnetosomes (soma = body), which are oriented parallel to the long axis of the cell. A fascinating aspect of these organisms is that most species of bacteria from the southern hemisphere swim toward the south magnetic pole, while species naturally occurring north of the equator tend to seek a northward-facing path. Still other species don't have a preferred north- or south orientation but can swim in either direction along the magnetic field.

So, what is the *purpose* of magnetotaxis? Life needs oxygen or some other element that can serve the same purpose, namely to act as an electron acceptor during the production of energy. Some organisms require lots of oxygen; others can do with small amounts; still others are actually harmed by it (see the story on the bacteria that cause botulism and tetanus). It seems that magnetotactic bacteria use the magnetic field for "up" and "down" orientation until they find the optimum oxygen concentrations in water.

Observations that these bacteria are typically found at certain depth zones in bodies of water gave the first clues that they prefer either oxygen-free environments, or that they need very low concentrations of oxygen in their surroundings. According to Richard Frankel, a biophysicist at California Polytechnic Institute who studied the mechanism extensively, "magnetotaxis increases the efficiency of aerotaxis in vertical concentration gradients by reducing the three-dimensional search problem to one dimension." In other words, magnetism provides a road map to find and stay at preferred oxygen levels.

I always caution my students not to call any organism "primitive." The word has a negative connotation that life does not deserve. In its place, I suggest that they may use the word "simple" to characterize a particular life form. Even this word is inadequate, for there is nothing simple or unsophisticated about bacteria. Just ask a member of the genus *Magnetospirillum* as it zooms by you toward its comfort zone.

# A HORSE OF A DIFFERENT COLOR

*Natural selection acts only by the preservation and
accumulation of small inherited modifications, each
profitable to the preserved being.*
*-Charles Darwin*

One of the best documented histories in the evolution of an animal is that of the horse. That's mostly because there is a rich trail of fossil evidence to chronicle its changes over time. People are often surprised that the horse – or its ancestors – originated in North America because it eventually disappeared from this continent until the reintroduction of the modern horse by the Spanish in the 15th century.

Ever since naturalists dared to commit the heresy to suggest that the Earth is more than a few thousands of years old, the idea that some creatures lived and died out a long time ago, while others underwent many changes in form and structure, took hold. In the 19th century a new science of paleontology was born whose disciples began digging and examining fossils in a systematic way. It quickly became clear to geologists that some rock layers were very old and contained remnants of organisms that disappeared from younger rocks. They vanished forever. They became extinct. Comparative anatomists began to explore some of the relationships between earlier and later forms and offered

118

possible explanations for the links between the different physical characteristics.

The ancestors of the horse can be traced back to about 50 million years ago. These animals did not much look like the modern horse: they were small (variously described in the literature as a small dog, or the size of a deer; more about the size discrepancy later); their teeth were unsuited for eating grasses; and they had four padded toes on each of their front feet and three on each hind foot. This "dawn horse" (eohippus) probably lived in forested areas rather than on open plains, and consumed leaves and twigs. Biologists gave this early horse the name *Hyracotherium*, meaning "hyrax-like beast." It roamed the woodlands of western North America and, when land connections between North America and Eurasia existed, that continent as well. The first fossil was discovered in England by paleontologist Richard Owen, who mistook it for a hyrax-like animal, hence the name. Specimens later turned up in many localities in what are now the United States and Canada.

There is a lot of confusion in the literature regarding the sizes of these early forms. Some writers mention a "very small dog" size, while others refer to it as a "deer-like" creature. Recent (2012) evidence indicates that the size of *Sifrhippus* (yes, it has been renamed!) fluctuated with changing climates. Within a couple of hundred thousand years, as global temperatures increased, average body sizes decreased. As the climate began to cool, the body sizes of the animals increased. This is a well-known phenomenon to ecologists. Documented in many organisms, it is called

Bergmann's Rule, and is apparently the result of the smaller surface-to-volume ratio in larger animals requiring less energy to maintain steady body heat.

As the climate changed, so did other aspects of the animals' bodies. Cooler, more temperate climates led to the development of grasslands. The food changed, as did the landscape. Tough, fibrous grasses require a very different set of teeth than do leaves and other foliage found in forests. Open savannas no longer hid animals as forests once did: those animals that could outrun their predators had a selective breeding advantage. Thus, the horse gradually changed. Teeth became more suitable for grinding cellulose fibers. Longer legs were better suited for running in open land. Padded feet gave way to more efficient hooves for running on hard surfaces. Toes became less numerous – eventually reduced to a single middle toe. The others were "lost", though they are still present in the early developmental stages (actually, the modern adult horse retains a single, useless vestigial toe, called a splint).

All of these changes are well documented in the geologic record. Intermediate forms from about 35 million years ago (*Mesohippus;* 3 front toes), 20 million years ago (*Merychippus;* still 3 front toes but the two side ones already greatly reduced in size); 5 million years ago (*Pliohippus;* single toe) have been preserved in the rock strata of North America, Europe, and Asia.

The modern hose, *Equus,* appears in the so-called Late Pleistocene. Prehistoric artists recorded modern horses in their paintings some 25,000 years ago. Discovered on the walls of caves in France, the illustrations included images of dappled horses. Was this "artistic license" or did these horses actually have spots? Newly available DNA data indicate that some of these horses indeed had leopard-spot patterns.

However, after the most recent ice age, the horse became extinct in North America and disappeared from Europe. It survived in Asia, from where it was re-introduced into Europe by marauding eastern tribes within recent recorded history.

Students entering into my freshman biology classes think of evolution in vague terms of "organisms adapting themselves" to changing environments. In survey after survey that I administer on the first day of class (please see the next essay), more than half of the students write something like that in response to my asking them for a definition of evolution. It takes months of talking about genetics and other areas of biology before we can undo this spurious concept.

Organisms cannot, of course, change themselves to adapt to changing environments. They can, however, be born with certain traits that allow them to survive, if that trait confers some advantage over others in the population. This is the basic tenet of what biologists call "modern synthesis": a marriage between the science of genetics and the Darwinian concept of "descent with modification."

The evolution of the horse is a good example. The ancestral horse did not simply wake up one morning to decide over a breakfast of tasty leaves and twigs that fewer toes and stronger teeth would serve it better from then on. For the horse, it was the trial-and-error of genetic mutations and natural selection that led to the optimum single-toed condition best suited for running.

It is also important to remember that evolution is not always linear. In other words, several different "versions" of an animal may live at the same time. Which "model" will prevail over time is dependent on the environmental conditions of the area. People who criticize evolutionary theories often point out that there are several forms of an animal co-existing at the same time. In their "fixed species" concept, they have a difficult time accepting that nature could be experimenting with different forms.

For many people, explanations of evolutionary changes are hard to swallow. Some are simply incredulous that drastic change over time is possible. Some have a hard time with the seeming randomness of it all. Some are possibly offended by any inference to human evolution and a nonhuman primate connection. Still others may reject evolution for religious reasons.

Yet, as the famous geneticist, Theo Dobzhansky, proclaimed: "Nothing in biology makes sense except in the light of evolution."

# "LITTLE MONKEYS ON THE GRASS.."

*Chimpanzees didn't prepare the way to humans*
*any more than we prepared the way to them.*
*-Joan Dunayer*

Opinion survey polls consistently show that of those Americans who have a stated position on the subject, about 40 percent accept the scientific concept of evolution. This, of course, also means that 60 percent of the public reject it. Yet, after many years of teaching both traditional-age and adult students in a variety of settings, I began to suspect that people on both sides of the issue have no clear idea what evolution is. Strong opinions are likely proffered about incorrect concepts.

To test this idea, for several years I polled freshmen students in my introductory college biology classes during the first week of the Fall semester, long before I spoke a single word about the subject of evolution. I wanted to see what concepts, knowledge, or dogma they brought with them from their high school days.

The survey I administered was in no way scientific, nor did I attempt to perform statistical analyses of the results. Yet, the outcome was quite eye-opening and the results were virtually identical from one year to the next.

The survey worked like this. I gave each student a sheet of paper with three questions. They were asked not to put their names on the paper. The questions were:

1.Please define the word "evolution" to the best of your ability;

2.Now that you wrote a definition, do you agree with this concept? (yes or no);

3.a. If you agree with YOUR definition of evolution, why?

or

b. If you disagree with YOUR definition of evolution, why?

I gave the students 5 minutes to think about their answers before they wrote anything. Then I asked a student to collect the surveys when everyone was done writing, and I left the room for about 10 minutes.

I collected a total of 306 surveys from seven sections over a four-year time span. As I began to read them, it quickly became apparent that most answers could be grouped based on similar themes, and I eventually sorted the papers into six categories. These were (including some "typical" answers – complete with spelling and grammatical errors – as points of illustration):

### 1. Organisms adapt themselves to their surroundings:

*"Evolution is the process caused by the organism needing to adapt to a changing environment."*

*"...the natural way that things change themselfs to adapt to their surrounding."*

One hundred twenty eight of the 306 respondents (42 percent) put answers that fell into this teleological category; 119 of the 128 agreed with their own statements. This theme was, by far, the most popular answer.

**2. Organisms go from 'simple' to 'complex' or 'advanced':**

*"The process in which living organisms became bigger, better, smarter with the passing of time."*

This definition was provided by 40 (13 percent) students. All but 8 students agreed with the statement.

**3. Changes occurred over long time:**

*"Over long periods of time new things will emerge due to the revolving time period also known as evolution."*
*"The development of any species over long periods of time. They could be small changes or extreme changes."*

This type of answer was the choice of 46 students. All but four of these individuals felt that this was a definition with which they agreed.

## 4. Apes to humans:

*"Evolution as I know it is the transformation of animals over time. Specifically Neanderthals into humans. I'm not sure how the classes or stages go but they include primates, neanderthals, cromates, and others I forget."*

Twenty-two of the 306 students answered something similar to this. Perhaps not surprisingly, 9 of the 22 respondents strongly disagreed with their own definition.

## 5. "Other" answers:

This category included a variety of answers that I was either unable to characterize, or that made no sense, or were ones that hit all previous concepts combined into one answer.

*"The word evolution to me means how the world has come about and new generations became what they are."*
*"Evolution could mean almost anything. The world, a person, plants even. I think anything can evolve into something. Or become evolutionized."*

A total of 54 (18 percent) answers were lumped into this group. Since these answers were either meaningless (see above) or confusing, only 34 of 54 students agreed with their own statement; 16 disagreed, and 4 marked both "agree" and "disagree" boxes. Clearly, these individuals were not only confusing but confused, as well.

## 6. More-or-less correct answers:

Later on I will define what evolutionary theory means to most biologists, but the answers in this category (16 students; or 5 percent of the survey respondents) came closest:

*"Evolution is a process through which a species passes their gene and its biological features to the next generation with some modifications due to the environment."*

Thirteen students agreed with their statement; one disagreed, and two both agreed (for all creatures other than humans) and disagreed (humans were created separately).

Among those who disagreed with whatever category of answers they provided, most stated religious reasons for doing so, even though I never asked them about their faith or religious orientation. Some of the typical reasons are cited below (again, complete with stylistic and spelling errors):

*"As a religious person, the Bible states God created the heavens and earth. It was written for Jews that had no scientific [knowledge] of the earth. Thus it would be observed that it would have included the information in Genesis. If God created the universe it seems to me that he could use any method he see fit."*
*"I believe that God created the animals the way he wanted them. I don't believe that dolphins became dogs or that we were once Apes."*

So, is this what we have been teaching them in our schools, homes, and houses of worship about evolution? If this is what the general public knows about evolutionary theory, is it any wonder that a majority of the American public doesn't "believe" it? Equally importantly, as even this non-scientific survey shows, even those individuals who "accept" the concept apparently don't have much of an idea about what they profess to be true.

I have thought long and hard about why there is this confusion and mistrust about what evolution means. First, I think the problem is one of semantics. The word "evolution" has different meanings in different disciplines. A quick check of the dictionary (Webster's New Universal Unabridged, 2001) provided no less than ten different definitions for the word. Included in these were references to "growth and personal development" (in the context of the social scientist's use for personal achievement); a "process of gradual, peaceful, progressive change or development" (the political scientist's use of the term); "movement of troops, ships, etc, for battle" (the military scientist's interpretation); and several other explanations.

Another troubling word is "theory." For the scientist, use of this term is restricted to the context of a rigorous process of hypothesis testing that includes evaluation of experimental data. For the general public, the word has quite a different meaning that can include guesses without any necessary knowledge of facts or analysis of empirical data. It is no wonder, then, that one often hears that "evolution is just a theory, not fact."

Secondly, evolutionary theory has been, at best, misunderstood, or simply painted as a canard. Scientists of all stripes – from Galileo to Darwin – have been branded as scoundrels, criminals or heretics if they dared to deviate from the official dogma of their respective era. The denunciation in the 17th century by the Catholic Church of the Copernican concept of a heliocentric system of planets; the Scopes "Monkey Trial" in 1925 in Tennessee; and the death of the Russian geneticist, Nikolai Vavilov, in a Soviet prison camp in 1943 for being "an enemy of the people," are among some of the most egregious examples of enforcing the official doctrine of their day.

While I'm not a conspiracy theorist at heart, I do believe that powerful forces always want to maintain the status quo, and they go to great lengths to confuse and misinform people. Obfuscation serves a purpose. After all, knowledge is power and power is not meant to be shared with others by the rulers.

Thirdly, our educational system and parents have failed our students. Some teachers have an inadequate knowledge and understanding of the subject. Others don't "believe" in it for their own personal reasons. Still others are reluctant to broach the topic for fear of stirring up controversy. In high schools, especially, there is no such thing as academic freedom. Parents, school boards, principals, and state legislatures often intimidate or threaten teachers into silence and/or demand the teaching of religion-based concepts regarding the origins of organisms.

Many teacher organizations fight back, of course, but this is often a very expensive and energy-intensive battle. The end-result of these obstacles to good science education is that students are confused and are left with some murky concepts of dinosaurs and ice ages that they acquired from watching Disney movies.

Lastly, people don't have the faith in science or the trust in scientists that were once widespread. A case in point is a Pew Research Center report, released in December, 2016, in which only 35 percent of adults say they fully trust scientists to give a full and accurate account of the health effects of genetically modified foods. And a recent revelation that Harvard scientists accepted bribes from the Sugar Research Foundation in the 1960s to implicate saturated fat, instead of sugar, in elevated heart disease risks, became the basis for recommendations by the government to curtail the intake of high-fat foods. The food industry, then, switched to a decades-long practice of high sugar content for enhanced flavoring of prepared foods and this may be, in part, responsible for the tremendous increases in obesity and diabetes mellitus type II cases in the U.S.

By the time I get around to explaining the basic tenets of evolutionary theory in my classes, we have spent a semester-and-a-half talking about chromosomes, genes, embryology, biochemistry, the cell theory (interestingly, a "theory" that no one seems to doubt anymore), taxonomy, and a host of other topics. In each of these subjects, I attempt to relate the topic to the theme of evolution.

Having had the luxury of time to set the stage properly, I finally get around to focusing on evolution for a period of about three weeks. However, before I begin, I tell my students: "I am not here to convince you of anything. If, at the end of this section you want to reject evolutionary theory for whatever reason, please feel free to do so. I ask only that you understand what it is you reject." I shouldn't really have to say this; after all, this caveat is never used in discussing the heart and blood circulation, or in describing reproduction in ferns, or for any other topic we deal with during the semester. At the same time, I do realize that this is a sensitive issue for some students.

So, how do I "define" evolution for my students? Like many biologists, I approach the subject initially from a population genetics point of view. I tell the class that while individual organisms are not capable of "evolving" or "adapting themselves," there are changes in the genetic make-up of populations over time. Different individuals are born with slightly different traits, some of which may be more advantageous to them if and when the environment changes (please see the previous essay on the evolution of the horse). I explain that the time period for evolution can be long or short, depending on the kinds of organisms and the environmental pressures that are exerted on them.

I make a deliberate distinction between genetic changes in populations "as a result of," rather than "in response to" changes in the environment. I try to clarify that "descent with modification" is not synonymous with "dolphins becoming dogs," or that "we were once apes."

My worry is not whether students "accept" the theory of evolution. What makes me appalled is that only about five percent of the students identified the process at least partially correctly in the first place. Most people have an opinion on the subject of evolution in survey polls, yet, apparently, very few know what they are agreeing or disagreeing with.

I test my classes on the scientific explanation for evolution, but I do not ask my students after my lectures whether they were "convinced" by my presentations, or whether they "changed their minds." I do not feel that that is my role, and I want to be respectful of their individual beliefs, whatever those may be. I do hope, however, that they come to their own conclusions and, if they don't agree with the scientific version of the "theory of evolution", then, at least, they know what it is they are rejecting.

And I always hope against hope that I will never, ever again see a response like this:

*"If evolution theory or concept or whatever is true then why don't we keep evolving? ...If the evolution theory is true for humans then why aren't little monkeys sitting outside on the grass evolving into humans?"*

\*\*\*\*\*\*\*\*\*\*

This essay was first published in a slightly different format in 2010, as a short article in the journal *Evolution Education & Outreach,* Volume 3, pages 416-419.

# "ONE IF BY LAND..."

*...the whiteness which surprises you is
caused only by the presence of myriads of infusioria,
a sort of luminous little worm, gelatinous and without
color, of the thickness of a hair, and whose length
is no more than seven-thousandth of an inch.
These insects adhere to one another sometimes for
several leagues under the sea..*
*-Jules Verne*

In the Disney movie, *Finding Nemo*, Marlin, the clown fish and his accidental but well-intentioned companion, Dory, are on a mission to find Marlin's kidnapped son. It is a journey fraught with peril – from sharks to humans, danger lurks with every ripple of water. Along the way they encounter an angler fish, whose illuminated lantern mesmerizes our heroes. This almost costs them their lives because the luminescence is attached to a mouthful of needle-pointed teeth.

The angler fish appears to be a popular villain for its ferocity and its ability to generate light in the cold and dark abyss. It is featured in *The SpongeBob Squarepants Movie* and in a dozen or so video games. It is one of hundreds of seafaring creatures that can generate a strange and mysterious glow in an inhospitable environment. This phenomenon is known to science as bioluminescence.

People have been fascinated by this curious biological light-producing process encountered in water and on land since recorded history, and probably well before they started to write things down. Aristotle pondered the nature of bioluminescence, and the Roman philosopher, Pliny the Elder, mentions it in his writings. And, of course, many of us have fond childhood summertime memories of lightning bugs.

One of the key concepts in biology is that complicated biochemical processes are solved once in evolution. The components of the nucleic acids DNA and RNA, for example, are basically the same from viruses to vertebrates.

Yet, many biochemists suggest that bioluminescence has evolved dozens of times among widely different groups of aquatic and terrestrial organisms. If so, then the process of generating light must not be all that complicated. A possible alternative explanation is that light serves critical functions in the lives of those creatures that make it. Both explanations have merit and both may be correct in some cases.

Physics and chemistry books are quick to point out that bioluminescence should not be confused with other forms of illumination such as fluorescence or phosphorescence, or the generation of light by incandescent fixtures made by humans. Rather, bioluminescence involves a chemical reaction that emits "cold light." In that respect, it most closely resembles the familiar glow stick used for recreational and military purposes where two chemicals are mixed together in order to generate light energy.

In living systems, the chemical process involves the oxidation of a molecule called luciferin (whoever came up with the name obviously had a sense of humor). The reaction is speeded up by an enzyme named luciferase and requires oxygen or some other oxidizing agent, as well as a co-factor (usually calcium or magnesium). There are many variants of the enzyme and many more are likely to be discovered in the future.

Bioluminescence is most frequently encountered in the ocean – especially in very deep and dark waters. The reason that we are just now beginning to appreciate how common this phenomenon is in the sea is obvious: until recently there was no way to descend and look at what's happening way down there. With today's remotely-operated submersibles and superior camera technology that is beginning to change.

On land, it has always been easier to observe organisms that glow in the dark. Among terrestrial creatures, some insects or their larvae, certain bacteria, and a handful of fungi can light up. In a 1526 A.D. publication on the natural history of the West Indies, the author notes that Spanish soldiers attached small pieces of luminous rotted wood to their helmets to prevent separation of the troops from one another and getting lost during nighttime maneuvers. And the 18th century American preacher and philosopher, Cotton Mather, makes a reference in 1721 to a book written by a Swedish author in 1555: "Olaus Magnus admires the Benefits which the rotten Barks of Oaks give to the Northern People, by the Shine, with which they do in the long Nights direct the Traveller." In fact, there are a number of fungi that glow in the dark. One of these fungi is *Armillaria*, a wood-rotting fungus about which I wrote in the essay entitled "Bigger Than a Breadbox."

If an organism makes certain chemicals to generate light, it most likely derives some gain from the expenditure of energy. Some of the benefits are fairly obvious and are well documented. The functions of light in marine creatures are usually divided into two main categories: defensive mechanisms and offensive purposes.

Among the list of defensive measures are: to startle an enemy; to confuse a predator; to distract an adversary long enough to make a quick getaway; to warn an opponent; or to "point the finger" at someone. An example of the latter is the use of light by microscopic organisms, called phytoplankton, when fed upon by small crustaceans. When the predator attacks a swarm of phytoplankton, the turbulence in the water causes the prey to light up like a motion-detector light. This, in turn, lets the predators of the crustaceans know that there is a meal for them.

Light also can be used for offensive purposes, such as to lure prey (like the angler fish mentioned above); to disorient or confuse a potential prey; or simply to light up a meal. Light also may be used as a means of communication and in finding a potential mate. Marine bacteria may produce light in order to *be* consumed. In a series of recent experiments, reported in the *Proceedings of the National Academy of Sciences*, the bioluminescent bacterium, *Photobacterium leiognathi*, has been shown to be eaten by microscopic organisms, called zooplankton, which then begin to glow. The plankton, in turn, are easily found and consumed by fish. But the bacterium survives in both the zooplankton and in the digestive system of the fish. When the fish defecate, they release the bacterium into new environments and this affords an effective dispersal mechanism for the microbe.

On land, organisms also may use light for different reasons. For the firefly or lightning bug, light is a means of finding a mate. Different species of fireflies emit different patterns of light pulses to signal their own kind that they are available for mating. The frequency of light pulses generated is controlled by the insect through its intake of oxygen. Along the side of the abdomen are a series of openings, called spiracles, that serve as breathing mechanisms for the animal. By regulating the ebb and flow of oxygen, they also are able to control the chemical oxidation process of luciferin and the pattern of luminescence. This means of communication has to be used carefully because predators are also watching; an overly amorous and careless individual may become a meal instead of finding nuptial bliss.

And what benefit do wood rotting and other fungi derive from illumination? The answer is not clear-cut, but one hypothesis is that insects and other invertebrates are attracted to the light and pick up the spores of their illuminated host and thus aid in spreading the fungus to new areas.

In the blockbuster motion picture, *Avatar*, director James Cameron uses bioluminescence extensively in telling his story of the fictional planet of Pandora. The use of light is not only a delicate and surreal artistic ploy, giving many of the scenes a beautiful and ethereal feel to contrast with the depiction of avaricious human exploitation of the planet's natural resources, but it is also a symbol of universal connection between living things.

Humans have always been fascinated by light. Fires meant not only warmth and a means of rendering foods safer and easier to digest, but also allowed us to see in the dark. Our poor eyes, lacking the shiny tapetum lucidum layer that most other mammals possess, prevented us from peering into darkness where so much real or imagined danger waited for us. In Greek mythology, Prometheus steals fire from Mount Olympus and gives it to humans. It is an act for which Prometheus is punished mightily by Zeus, but people finally have light with which to see in the darkness. For thousands of years, humanity used a variety of fuel sources to chase away the darkness that terrifies us so. Modern electric light fixtures have profoundly changed our daily lives in the last 130 years or so.

On the other hand, many organisms, including plants, fungi and "lowly" animals have been using light for lots of purposes for tens of millions of years. This, yet again, is a good example of how living systems have solved complex engineering feats long before *Homo sapiens* came along.

# A STICKY SITUATION

*Any living cell carries with it the experience*
*of a billion years of experimentation by its ancestors.*
*-Max Delbrück*

Newtonian gravity is a great thing. It keeps my car on the road and the dinner plates from floating off the dining room table. Sometimes, though, gravity is not helpful: lifting 50 pounds of cracked corn from the truck or hanging a picture on the wall come to mind.

Such are my brilliant moments of soliloquy as I'm immersed in the mundane but unpleasant task of picking slugs in the garden. I think of gravity because these creatures have no trouble slithering along on the leaves of lettuce, or shimmying up the tomato vine or marigold stem. No wonder – slugs are covered with a film of sticky goo which sticks like orange-colored glue to my fingers. I confess that I greatly dislike these animals. It's not just that they are eating MY garden, but I find their soft, slimy bodies disgusting. Maybe, somewhere some gastropodologist – or whatever snail and slug experts are called – loves them, but most people I know loathe even the sight of them.

I am probably unfair to this houseless snail. Aside from the philosophical point that they have as much "right" to live as any other creature, including some other undesirables, like mosquitos, tapeworms, and bed bugs, slugs are complex invertebrates and the mucus they produce is a remarkable adhesive. Chemically, it is a complex polysaccharide material made up of long chains of sugar molecules combined with protein.

Slugs use this mucus for protection as well as movement and attachment to surfaces. The protective aspect is due to the very slipperiness of the body: just as I'm likely to drop a slug that I picked off a leaf, a potential predator may have a hard time hanging onto its prey. As for movement, the mucus provides a slippery track to allow for both locomotion and adhesion to objects.

It takes coordinated nervous and muscular action to generate the rhythmic pulses by which the body of the slug is propelled forward. The mucus is produced by specialized cells in the body, and the slug can control when and how the substance is released. Until needed, the polymer is stored in specialized little packets, called vesicles, inside the cells that produce it. This is important because the material is highly hygroscopic. Once released into the environment, it rapidly absorbs available moisture and swells up to 100 times its original volume, creating slippery, yet sticky strands.

The slime also partly protects the animal from desiccation, although slugs still don't do well under dry conditions. This is why they are mostly seen in the dew of the early morning or at sunset, or during rainy days. The goo has yet another benefit for the slug: they can hang onto each other during copulation. Each slug has both sets of sex organs (they are true hermaphrodites) but two individuals must cross-fertilize in order to produce new slugs.

Lots of other animals and plants produce glue-like substances during their lifetimes. Some marine worms construct their homes by building sand castles (well, OK, sand tunnels). This involves the production of two different kinds of proteins which must be bonded together quickly in a watery environment. Simultaneously released from separate specialized ducts, the components are solidified within 30 seconds. Sounds familiar? Crazy glue comes to mind, except that this bonding of sand and glue occurs under water, without dilution or loss of materials.

Insects are also well-known adhesive producers. They employ adhesives for building nests, setting traps, or to fasten their egg masses to plant surfaces like leaves or stems which provide a ready source of food for emerging the offspring. Some insects can even overcome the Teflon-like repellent cuticle of their plant host and actually incorporate their glue into the plant's outer cover to prevent the eggs from sliding off.

Humans discovered a long time ago that these materials can be used to hold things together or to counteract the pulling power of gravity. Hooves, hides and bones from terrestrial animals and fish scales are natural sources of adhesives. I remember my grandfather using glue to repair picture frames or fix chair backs and other loose or broken pieces of furniture. He kept the glue in a tin coffee can with the brush stuck in the dried goo. When he needed it, he would gently heat the glue until it liquefied. I will never forget the smell of this melting glue. When I asked him where this brownish stuff came from, he told me that they made it from horses. I was mortified that a pretty horse became glue, but he assured me that bones were used only after the animal died.

Plants, too, have been used for thousands of years for making glue. Starch and pitch are two substances that readily come to mind. Archeologists have uncovered numerous thousands-of-years-old pots that had been repaired using plant resins. Some plants release gums or pitch in response to wounding of the bark. Conifers use these substances to keep out bacteria and fungi as well as to trap insects and prevent their entry into vital tissues. Yet other plants actually produce sticky, sugary sap to attract certain kinds of bacteria or insects that keep away potentially pathogenic or destructive creatures.

The relatively recent realization that bacteria also make sticky substances has been an important scientific discovery. Many different kinds of bacteria appear to produce so-called biofilms – combinations of cellular appendages and slime by which they cling to one another as well as to living and non-living surfaces.

The implications of this discovery are considerable. Take fruits and vegetables, for example. Many disease-causing organisms, such as a potentially deadly form of *E. coli* cells can be present on the surfaces of these commodities. Yet, for most of us, washing fruit consists of a cursory rinse under a flow of tepid water. Does that really remove bacteria? Unlikely.

Or take another daily ritual, that of brushing of one's teeth. Do the toothbrush and toothpaste do a good job in getting rid of plaque-forming microbes? Or do we simply feel better about having done the "hygiene thing?" Actually, the fluoride in the toothpaste or rinse is a far more effective agent in killing bacteria than our feeble attempts at removing and destroying the oral flora by brushing.

Some scientists speculate that more than half of human bacterial infections involve biofilms. In addition to tooth decay and gingivitis, they appear to be involved in urinary tract infections, endocarditis, middle-ear infections and contamination of prostheses and other implanted devices. Biofilms are common on inanimate surfaces and represent a significant source of infection for hospitalized patients.

Bacterial cells produce the glue-like substances in a complex and step-wise fashion. Cells that are capable of becoming part of a biofilm colony first undergo a genetic transformation. Certain genes are switched on to produce the sticky material, called extracellular polymeric substance (EPS). This substance consists of long chains of sugars (polysaccharides) combined with proteins and DNA. (Sounds familiar? Apparently, both slugs and bacteria make very similar kinds of adhesives. One of the fascinating things about biology is how vastly different organisms solve similar problems in a similar fashion).

144

Within the biofilm, bacterial cells can communicate with one another via chemical signals. Cells can be added to the colony or cells may leave the group at some point. Some biofilms are composed of a single bacterial species, while others are combinations of several different kinds, and may include fungi, algae or other unicellular organisms. Clustering and cooperation among the cells can offer a lot of protective and nutritional benefits for the membership and also prevents easy dislodging of cells by external forces, such as washing of fruit or the use of antiseptic rinses.

Not all biofilms are bad. The recognition of biofilm clusters has led to the development of industrial applications, such as sewage treatment, or cleanup of oil spills in waterways. Much research is now focused on generating electricity from waste products and previously discarded biomass by using biofilm technology. So, once again, we see that many organisms have solved complicated problems which we humans are just now beginning to understand and hope to exploit in the future.

# RED WIGGLERS

*One of these fine days the public is going to wake up
and will pay for eggs, meats, vegetables, etc.,
according to how they were produced.*
*-J.I. Rodale, 1942*

Whenever I move from one place to another (and there have been a lot of moves over the years) one of the first things I do is to start a compost pile. Some of these have been as small as a plastic pail in a corner of the kitchen counter; others are as much as a meter in diameter and constructed of rot-resistant wood, wire mesh and hinged removable sides. I compost for many reasons. For one, I hate to throw food away but some parts of fruits and vegetables are not edible. Then there is the spoiled food despite my best efforts to keep that to a minimum. I also like to take the finished product of composting and use it as fertilizer in my garden. I think of it as a recycling program for nutrient molecules.

Surprisingly, I have been able to have gardens even in large urban areas. Sometimes these were tiny patches of ground in the back or front yard or even wooden boxes of soil on balconies. Now that I live in the countryside and have a few acres of land, I have a fairly sizable garden. Raised beds are an absolute necessity in a country where our best natural crop is rocks (see the story "A Stone, A Windowpane and the Nightly News" elsewhere in the book) in a heavy

146

clayey soil that is alternatingly waterlogged in early spring or harder than concrete in late summer.

The raised beds are filled with good soil, aged manure from the chicken house or cow manure from a nearby farm and, of course, composted stuff from the kitchen. I compost everything that doesn't have meat, dairy products or lots of fat in it. Vegetable scraps, coffee grounds, eggshells, grass clippings, small twigs and leaves, ashes from the fireplace, straw from the chicken house, biodegradable packaging, and moldy or spoiled food of all kinds go into the bin. I turn the pile at least once a week when it's not frozen to make sure that it is well aerated.

Composting as a means of amending the soil with nutrients is as old as agriculture itself. For thousands of years peasants revitalized soil with manure, crop plant debris, animal bedding, fish entrails, ashes from heating and cooking fires and inedible parts of foodstuffs. Organic methods of farming were the norm until about 100 years ago when chemical supplements began to be used in ever larger quantities.

It is interesting to note that the agrochemical methods are now considered to be "conventional" agriculture while the organic approach is called the "alternative" means of farming. Isn't this backwards terminology? Isn't there something wrong with this picture?

Even though modern agriculture relies heavily on chemicals and mechanization, not everyone agrees with these methods of producing food. The late 20th century witnessed a resurgence of the ideals of organic farming and locally produced foods. One of the most influential – and controversial – pioneers in the field (pun intended) of organic farming was Rudolf Steiner. A philosopher by training, Steiner became a vocal proponent of "biological dynamic farming" or, simply, biodynamics. He laid out his vision on how to "influence organic life on earth through cosmic and terrestrial forces" in a series of lectures in 1924. Although often heralded as the first "holistic" organic methods of farming, his ideas basically incorporated methods that have been practiced for thousands of years: integration of crop planting with animal husbandry; the use of composted plant materials to replenish soil nutrients; the cultivation of indigenous plants and the use of locally grown crop varieties; crop rotation; and the use of astronomical charts in the timing of planting and harvest.

Steiner's biodynamic methods have a dedicated corps of followers around the world. Others, however, ridicule his inclusion of cosmic and spiritual forces into his concepts. A frequent criticism of his work is that the "subjective and mystical approach" cannot be tested and verified scientifically in an attempt to compare the efficacy of biodynamic agriculture with so-called "conventional" (i.e. agrochemical) techniques.

Some of Steiner's techniques *are* somewhat mystical, bizarre and ritualistic. His insistence on packing manure into cow horns (preparation Number 500) to channel cosmic energy; stuffing powdered quartz into cow horns (preparation

Number 501) prior to burying it underground during the summer; the use of homeopathic quantities of herb blossoms, stinging nettle, oak bark and horsetail plants packed into cattle intestines or stuffed into skulls, urinary bladders or other body parts of different species of animals; the occult methods prescribed for applying the dissolved plant parts; and using the phases of the moon to guide planting and harvesting schedules, are examples of what have been characterized as pseudoscience and alchemy.

Rudolf Steiner did not help his own cause by insisting that proof of effectiveness was not needed since all of his recommendations were axiomatic truths and should be accepted at face value. Steiner may have been very confident in his beliefs, or he was supremely arrogant in his insistence of prima facie acceptance of his values. While he may be the most famous and often maligned advocate of organic agriculture, he is not its only proponent. The late 19th century British agronomist, Sir Albert Howard, and the more contemporary American organic farmer, Jerome Rodale, were influential workers whose sound scientific, data-driven methods have avoided the kind of criticism leveled at Steiner.

Key to healthy soil in "sustainable" growing methods is the earthworm. From the records of Egyptian pharaohs to the practices of modern organic agriculture, the utility of earthworms has been recognized for thousands of years. Charles Darwin, that precocious naturalist who dabbled in everything from beetles to plants to earthworms, wrote in 1881: "It may be doubted whether there are many animals which have played so important a part in the history of the world as have these lowly creatures."

The case for earthworms is eloquently stated in Peter Tompkins' and Christopher Bird's book, *Secrets of the Soil, New Age Solutions for Restoring Our Planet* (1989) as they devote many pages to singing the praises of worms and their contributions to soil building. They write: "Earthworms can produce more compost, in a shorter time, with less effort, than any other method."

Not only do worms process large amounts of soil through their digestive systems, thereby greatly improving the porosity and texture of soil, but their excreta, called "castings," enhance it with valuable natural nutrients.

Every time I turn over my compost pile I see huge numbers of worms. I have often wondered where they come from because I never bought any (they are available for purchase) to put into any of my compost bins. A bit of literature search has revealed that the "compost worm," also known as the "manure worm" or the "red wiggler," arrives with aged cow manure. Known by taxonomists as *Eisenia fetida*, the worm is a prolific inhabitant of high quality compost piles. And not just any earthworm will do: of the hundreds of species, only the red wiggler is an important resident of compost.

Just how prodigious are these worms? How many of them are in the compost pile? At this point, I have to confess something yet again: I am a bona fide science nerd. After marveling for months at the writhing masses of worms I see with every pitchforkful of compost, I decided to count them and come up with some numerical evaluation to try to answer those questions.

Rooting through several cubic meters of compost was out of the question. But a sample seemed possible. Accordingly, one lovely autumn morning, I filled a 29 cm x 15 cm x 10 cm (11.25 in x 6 in x 4 in) shoebox with composted matter and sat down at the picnic table to count worms. In that 4,350 cubic cm (270 cubic inches) volume of the shoebox, there were 573 worms of all sizes. And I'm quite sure that despite hours of patiently sifting through the compost and methodical counting, I missed at least 10 percent of them, especially the tiny baby worms that were barely visible to the naked eye. I have never been great at math, but my current compost bin is about 100 cm x 100 cm x 80 cm (800,000 cubic cm) in size. So, there must be a gazillion worms in there (well, dear reader, maybe you can calculate a more accurate number).

Further reading on the subject of worms told me a lot about these annelids and their use in compost-based agriculture around the world. According to a Canadian publication, *Manual of On-Farm Vermicomposting and Vermiculture*, "...in 2003, an estimated one million tonnes of vermicompost were produced on the island [of Cuba]. In India, an estimated 200,000 farmers practice vermicomposting and one network of 10,000 farmers produces 50,000 tonnes of vermicompost every month."

Perhaps Ann Raver's comment is best when she says in her essay, "The Dirt on Earthworms," in her 1995 book, *Deep in the Green*, that earthworms can turn hard, concrete-like ground into "black gold." It is clear that earthworms are as crucial to proper soil management as are bees for the pollination of many of our most important crop plants.

# BIGGER THAN A BREADBOX

*There is something fascinating about science. One gets such wholesome return of conjecture out of such a trifling of investment of fact.*
*-Mark Twain*

Folks in the news business love sensational stories. I guess that's because we, the consumers of news, like the bombastic stories. However scant or preliminary the information may be, it is quickly turned into "fact," truth be damned. We can always make corrections later. As the old saying goes, it's always easier to apologize later than to check for accuracy up front.

This is particularly true for stories concerning human health. One egregious example is a television commercial that touts the benefits of a certain cereal as "*some* scientific studies *suggest* that ___ (product) *may reduce the likelihood* of certain types of ___ (disease) under *certain* conditions when used regularly." And based on this gobbledygook I should buy this product? Some of us may remember the extract from apricot pits that was supposed to cure cancer. How about "colon cleanser" teas that remove accumulated deposits (?) from the intestines? Or the watermelon diet that promised quick weight loss? This nonsense is not restricted to foods or health products: lately, there has been a lot of talk of "clean coal" energy. Now, there is an oxymoron.

153

Part of the reason that these outrageous claims are perpetuated as fact is that we WANT to believe these stories to improve our lives, health and comfort without having to examine the details. Sometimes, we simply want to excite our senses and imagination by accepting things at face value. Even when science stories are more-or-less accurately reported, they are often sensationalized or distorted. Finding and re-reading a stack of old, yellowed newspapers the other day, I thought of all of this, and news of a "giant creature" that fascinated the public about 25 years ago. No, it wasn't Bigfoot or Sasquatch. It was a fungus.

Walking in the patch of forested land that I bought a few years ago, I see quite a few dead trees. Some were felled by loggers about 15 years ago and left there; others are still standing with dead branches and bark that's beginning to slough off. A closer examination of the stumps and bases of trees reveals a network of black, stringy structures spreading up the trunk from the forest floor like crudely spun spider webs. I recognize these as part of the body of a fungus that is called the "shoestring fungus," or the "honey mushroom." It is known to science as *Armillaria mellea*, a wood rot organism that can get nutrients from dead wood, or can live as a parasite that eats and ultimately kills living trees.

The black shoestring-like structures are called rhizomorphs and the fungus uses these to spread from tree to tree. In late summer or fall, clumps of honey-colored edible mushrooms spring up here and there, giving the organism its other common name. I took a rough count, and it seems to me that about 20 percent of my trees have the parasite. There is not much I can do about this other than to try to remove dead trees and stumps that serve as food base for the fungus. Even this is not practical, as I don't have the equipment to cut, grind and remove dead trees or stumps. Nature will have to provide the answer by allowing certain trees to survive, while others may perish.

In 1992, a modest four-page paper appeared in the "Letters" section of the science magazine, *Nature*. Titled "The Fungus *Armillaria bulbosa* is among the largest and oldest living organisms," it reported on an "individual *Armillaria bulbosa* that occupies a minimum of 15 hectares [about 37 acres], weighs in excess of 10,000 kg [over 20,000 lbs], and has remained genetically stable for more than 1,500 years."

On April 2, 1992, on the same day that the *Nature* article appeared in print, the *New York Times* ran a front-page story titled, "World's Biggest, Oldest Organism: Twin Crowns for 30-Acre Fungus." The *Times* may have calmly summarized the findings of the study but other news outlets weren't so restrained. Newspapers large and small, in North America and around the world, picked up the story and alluded to a giant ogre that menaced a large area in the Upper Peninsula of Michigan.

CNN wanted to send an airplane to take aerial footage of the site where the "humongous fungus" was "visible from the air" (actually, it was a clump of dead and dying trees that marked the site). The "pulsating mass," as one person described it, even made Dave Letterman's "Top Ten List." In many ways, the hysteria surrounding these scientific findings was reminiscent of the *Pfiesteria* story that is described in the essay, "The Cell from Hell," elsewhere in this book.

The city of Crystal Falls, Michigan, has started an annual "Humungus (sic) Fungus Fest," complete with duck races (?), a fungus-fest ball race (?) and, of course, the " humungus (sic) fungus pizza," which must be hoisted with a hydraulic lift to be baked.

Although Crystal Falls may have been the first to exploit the story for its financial benefit, even larger fungi were soon discovered. In May, 1992, a larger cousin of the shoestring fungus, *Armillaria ostoyae,* was found on a 1,500 acre site in Washington state. The news media feeding frenzy – having just calmed down from the April reports – started all over again: "Humongous Fungus Has BIG Brother Out West." This finding was surpassed, yet again, by a 2,200 acre behemoth, estimated to be over 2,400 years old (and maybe as old as 8,000 years), discovered in the Malheur National Forest in Oregon in the year 2,000.

At this point, you might ask, just how did the scientists know for sure that they were dealing with one very large organism spread over a wide area. The answer to that question is mundane and by now routine: laboratory pairings of samples, genetic markers, and DNA "fingerprinting" were used to reach this conclusion.

Tourists who visit Crystal Falls are reportedly disappointed. After all the hype, perhaps they expected to see an enormous toadstool. Or a huge, whale-like creature thrashing about. Or maybe something like a giant redwood tree, albeit at ground level. I don't know. But I do know this: this fungus is still sucking the life out of my maples, ashes, spruce, and other trees. I wish it weren't here in my woods and I'm not impressed by how big it can become.

# NATURE'S ALCHEMISTS

*Food is simply sunlight in cold storage.*
*The function of the plant is to store the energy*
*borrowed from the sun, preparing it for the*
*service of man and animals.*
*-J.H. Kellogg*

It is a beautiful late summer day and I am standing in the middle of a 15-acre woodlot with eight students. We are here to look at trees and how the forest regenerates after selective logging. Earlier in the day I laid out a transect line that we will use for the survey, and now we scatter in groups of two to take measurements. Now-and-then, one of us notices an interesting mushroom or insect, and we put our clipboards aside and reassemble in a clearing under an azure-blue sky and brilliant sunshine to look these over.

Most of the students in the group are not science majors, and most have not had college biology or botany courses. Five of the students are from major urban areas, but even the country kids don't know the difference between conifer species; they are all "pines," as far as they are concerned. I am aware of these limitations, and in this introductory forestry class we start with the basics: how plants grow, what they need in order to live, why some grow only in certain areas, and what plants mean to human societies. The students are sort of aware that all foods and lumber ultimately come from plants, but this knowledge is nebulous and they haven't really thought about this too much in their formative years.

Back in the classroom, sooner or later I am going to have to talk about photosynthesis. I always struggle with how much detail to put into my lectures. Do they really need to know the intricate details of this very complex process? Do I ask them to memorize every intermediate step, every enzyme, every chemical reaction? I tailor the amount of detail to different audiences, but my clue is to read the students' faces. It is uncanny how the same vacant look appears on all of their faces at the instant I lose them.

Most of us had to memorize in high school biology classes that photosynthesis is a process that plants carry out. The key ingredients, we were told, are water, carbon dioxide and light, and the products are sugars and oxygen. We dutifully learned these pieces of information, even though we never really understood all of it. And because we didn't understand it, and because plants are not nearly as exciting as kittens or puppies or exotic animals in the zoo, we never appreciated the importance of photosynthesis in our daily lives.

Well, guess what? Our high school teachers probably also didn't fully understand what they were teaching. Not only that, scientists only recently solved the puzzle of how plants, algae, and photosynthetic bacteria are able to carry out the incredibly complex chemical reactions involved in photosynthesis. Understanding it, however, does not mean that we can duplicate it. More about that later.

Life on Earth is possible because of a unique set of circumstances. In our solar system, Earth is just the right distance from the Sun to make liquid water possible. This is crucial for all living organisms for several reasons. First, water is very good at absorbing heat slowly and then releasing it just as slowly. This moderates temperatures on Earth for life as we know it.

Secondly, water is a major component of the cells and bodies of all living creatures. It is the solvent for most of life's molecules and it is a key ingredient for that most marvelous of all chemical processes: photosynthesis. Water is also the source of oxygen upon which nearly all organisms depend for their survival.

Photosynthesis first appeared on Earth about three billion years ago. We know this from the geologic record. Prior to the release of oxygen into air by photosynthetic bacteria (also called cyanobacteria or blue-green algae), Earth had a "reducing" atmosphere, indicated by the lack of oxidized iron (rust) in ancient rock samples. Cyanobacteria may have been the ancestors of true algae and plants. Only after enough oxygen accumulated in air and water did animal life become possible.

Knowledge of what is involved in photosynthesis is very recent. In the 17th century, Belgian physician Johann Van Helmont proved Aristotle wrong by showing that increases in plant growth were not the result of conversion of soil into biomass, but somehow involved water. But Van Helmont did not know anything about the roles of carbon dioxide and oxygen.

The release of oxygen by plants was not demonstrated until the 18th century when Joseph Priestley discovered the gas as a byproduct of plant metabolism (the French chemist, Antoine Lavoisier actually named the element). However, details of this incredibly complex process were not worked out until the latter half of the 20th century. Painstakingly, step-by-step, researchers described the hundreds of chemical reactions taking place in photosynthetic cells.

It is mindboggling to think that such complicated chemical processes arose in ancient life forms billions of years ago. Yet, nature also had billions of years to tinker with, and improve upon, the details. The end result is that from relatively simple molecules, and harnessing the energy of the Sun, the sustenance of life is generated. Without the carbohydrates and oxygen made by photosynthetic organisms, other life forms would not be supported. But make no mistake: there is nothing altruistic about this association. Plants make the sugars and oxygen for their own needs. Animals simply take advantage of the food, and the oxygen we get from plants is the excess amount that they do not use for their own needs.

So, why should we care about the details of photosynthesis? After all, we know that it works; we know that plants can convert solar energy into chemical energy to feed us and provide wood and other raw materials upon which we rely for our daily lives. The answers to this question are both heuristic and practical. Humans are curious animals and we want to know how things work. If we understand something, we may have a better appreciation for the organisms that are able to carry out the process. Most importantly, if we understand the mysteries of the system, we can learn from it and, perhaps, even imitate it and "improve" it.

Without going into minute detail, let us look at how the starting ingredients of light, water and carbon dioxide are used by photosynthetic organisms, and how we might be able to mimic or duplicate nature's designs. The energy of visible light (from sunlight or, to a far lesser extent, from artificial light) is used to break apart water molecules. Composed of one oxygen and two hydrogen atoms, water molecules represent stored energy in the form of chemical bonds. Boiling does not break these internal bonds; it only disrupts the link between adjacent water molecules.

Plants have figured out how to capture light energy and use it to take apart water molecules in order to liberate the energy within. (Imagine if we could imitate the capability of photosynthetic organisms to fracture water molecules efficiently. The key word here is efficiency because our current engineering methods take more energy to accomplish this task than what we get out of it).

Ok, so light energy is used to split water and liberate electrons, which are forms of energy. The electrons are then used to convert carbon dioxide into sugars. Water's contribution doesn't end with the supply of energy. The two hydrogens become incorporated into the carbon dioxide to make sugar, and the oxygen is liberated into the air. This is the oxygen that all creatures, including plants, use in another complex chemical process called cellular respiration.

Well, that's about as simple as I know how to describe photosynthesis and still cover the basics. Seems to be fairly straightforward, right? The answer to that question is, of course, a resounding "no." Testament to the complexity of photosynthesis is that several Nobel Prizes were awarded to the scientists who eventually figured out the entire process.

Now that people understand what photosynthetic organisms are capable of, and how they do it, we should be able to mimic their efforts. Unfortunately, that is easier said than done. Scientists around the world are currently engaged in trying to emulate the process that even "simple" cyanobacteria and unicellular algae accomplish quickly and effortlessly.

Some researchers are focusing on compounds and mechanisms to capture light in a manner similar to plants and algae. Efforts are underway at Arizona State University, Penn State, the Tokyo Institute of Technology, and a handful of other universities to construct "artificial leaves," or molecules that absorb light and transfer its electrons to intermediate energy storage units. Many problems remain, as current artificial systems are inefficient and use very expensive and rare raw materials. And when electrons *are* liberated, they are not always able to be channeled to the manufacture of useful substances but, rather, lead to the generation of "waste" heat.

If or when it becomes feasible to mimic nature's most marvelous accomplishment, the outcome could be nothing short of revolutionary. Several benefits immediately come to mind. One, of course, is the production of more food. Since natural photosynthesis is in itself inefficient in that only about 1 to 2 percent of the available light is fixed into sugars, we may be able to improve upon the process.

Imagine how much more food could be produced, and people fed, if we could just double or triple photosynthetic efficiency (of course, this may open Pandora's box in terms of population increase, increases in use of fertilizers and fuel, problems in landscape alteration, etc.). Improving the process of photosynthesis is currently being investigated at the University of California, Berkeley, by altering the genetics of plants so that their enzymes are able to process light more efficiently.

Photosynthesis is often referred to as a "carbon-neutral" process. This means that for every molecule of carbon dioxide used, we get a molecule of carbon-containing product (sugar). Taking this concept one step farther, we could potentially use components of the photosynthetic pathway for carbon "sequestration." Thus, we could remove from air one of the major greenhouse gases that may be contributing to global climate change. Carbon dioxide is not only one of the major greenhouse gases, but each molecule persists in the atmosphere for an average 120 years.

Thirdly, if we can mimic parts of the photosynthetic process to break water efficiently, we could generate hydrogen as a fuel source. Although hydrogen has been touted as a major potential alternative to fossil fuels, at present its production and distribution are problematic and too costly. However, hydrogen may be part of the energy-alternative mix in the future if these problems are solved.

If not hydrogen, then the electrons harvested from water may be used for other energy-storage molecules. Nature does this by using a chemical compound called ATP as an energy source to drive the chemical reactions involved in photosynthesis. If we can come up with stable and environmentally friendly energy storage intermediates, then the current problem of how to store solar energy produced by photovoltaic systems may be solved. This would replace the need for the highly polluting carbon-based fossil fuels we currently use.

Photosynthetic organisms are truly nature's alchemists. Although the initial product may be glucose or other sugars, these can be converted into oils, fats, proteins, vitamins, complex carbohydrates, and hundreds of useful products in fruits, vegetables, fibers, and wood. These treasures are worth far more than the precious metals human alchemist tried and failed to make from lead, iron or copper in ancient times.

Students often think I am joking when I tell them that on their way to lunch after lecture they should go and hug a tree or bend down and kiss a blade of grass. I am serious when I make this suggestion.

Thoreau, in *Walden*, writes:

*"We are wont to forget that the sun looks on our cultivated fields and prairies and forests without distinction. They all reflect and absorb his rays alike, and the former makes but a small part of the glorious picture which he beholds in his daily course. In his view the earth is all equally cultivated like a garden. Therefore we should receive the benefit of his light and heat with a corresponding trust and magnanimity."*

Without the sun and photosynthetic organisms we simply wouldn't be here to be able to saunter over to the cafeteria after classes, breathing oxygen while we are walking to get our next meal.

# "ONE OF THE MOST USEFUL AND BEAUTIFUL TREES..."

...is how Harvard botanist Charles S. Sargent described the American chestnut in his *Silva of North America* in 1890:

*"No other tree grows so rapidly or to such great size on the dry gravelly hills of the northeastern states. Always beautiful with its massive trunk, its compact round-topped head, and slender dark leaves, in early summer, long after the flowers of its companions have disappeared, the chestnut covers itself with great masses of spikes of yellow flowers, and is then the most magnificent object of the sylvan landscape."*

With an estimated 3.5 BILLION standing trees over 200 million acres of forested land at the dawn of the 20th century, the American chestnut (*Castanea dentata*) was an important part of the forests of eastern North America. The range of the species extended along the Appalachian Mountains from southern Maine to north central Georgia, and westward to central Alabama, northern Mississippi, central Illinois, and southeastern Ontario. Chestnut trees occupied just about any suitable habitat, from dry ridgetops to moist bottom lands, but grew best on well-drained slopes at 2-4,000 foot elevations.

In its northern range the tree reached 50-80 feet in height, but in the southern Appalachians 100-foot-tall trees with trunks 10-feet in diameter were not uncommon. In 1897 Gifford Pinchot, the future first chief of the U.S. Forest Service, reported finding several American chestnuts in western North Carolina that were 120-feet tall and 13 feet in diameter. Some of these trees were estimated to be more than 500 years old.

Over the entire range of the species, chestnut made up about 25 percent of trees in hardwood forests, but in many areas it grew in almost pure stands. In the Cumberland Mountains of Kentucky, for example, American chestnut and chestnut oak made up 60 to 85 percent of the forest on dry ridge tops. In western Pennsylvania some of the original forests may have consisted of up to 70 percent chestnuts.

Because the chestnut is frost-sensitive, its numbers dwindled in the northern states, but it was found in abundance on the coast, and in glaciated inland areas in New England near the Atlantic Ocean. Henry David Thoreau wrote in 1854:

*"They also grew behind my house* [in Concord, Massachusetts] *and one large tree, which almost overshadowed it, was, when in flower, a bouquet which scented the whole neighborhood, but the squirrels and jays got most of its fruit; the last coming in flocks early in the morning and picking the nuts out of the burs before they fell, I relinquished these trees to them and visited more distant*

*woods composed wholly of chestnut. These nuts, as far as they went, were a good substitute for bread."*

The dominance of chestnut began to decline rapidly after 1904, the year that a deadly fungus disease, chestnut blight, was first described. The causal fungus, *Cryphonectria parasitica,* was imported from China on Asiatic chestnut seedlings or packing crates, probably in the 1890s, but was not associated with the death of American chestnut trees until it was found on dying trees in the Bronx Zoo.

In 1910, reports of diseased and dead chestnut trees filtered in from New Jersey, Massachusetts, Virginia and eastern Pennsylvania. By the 1950's more than 80 percent of standing American trees were dying or dead throughout the entire range of the species.

Once infected, the trees quickly die because sunken lesions, called cankers, girdle the stem and prevent movement of nutrients and water in the bark's vascular system. The fungus uses the bark to get nutrients and to reproduce by producing millions of asexual and sexual spores.

Most likely there were several reasons why this fungus, dubbed the "King of Pathogens" by forest disease expert Paul Manion, succeeded in killing trees so rapidly and efficiently. The almost contiguous line of chestnut trees in the eastern U.S. provided an ample supply of hosts. More importantly, American chestnuts had not co-evolved with this Asian pathogen, and the trees did not have resistance to the foreign invader.

The fungus is also very efficient at producing spores, both by sexual and asexual means. And the spores can stay alive for long periods of time in chestnut bark. Once, I found an old stem piece in a laboratory cabinet, labelled 1930-something. Nearly 60 years later, I was able to grow out the fungus from that old specimen. The pathogen also is able to survive in a more-or-less dormant state in several species of oaks, which are usually not killed by the infection.

The death of the chestnut caused major changes in the species composition of the eastern forests. Plant ecologist Catherine Keever wrote in 1953:

*"Since the chestnut is almost gone and will certainly never again be among the dominant trees of the...forests of the* [Blue Ridge Mountain] *region, the association is no longer oak-chestnut and should not continue to be called by that name."*

In many areas the openings created by the death of chestnut trees was quickly filled by other species. In some forests the dominance of oaks (red, white or chestnut oak) increased. In other areas, a mixture of oak and hickory took over. Still in other stands, so many chestnut trees died in such a short amount of time that the forest reverted to red maple, black birch, black cherry and locust. These species were eventually replaced by sugar maples, beech and basswood. Any chestnut that may have re-sprouted from old stumps remained small, understory shrub-like vegetation. Large areas became dominated by the attractive, but less valuable, yellow poplar (please see the essay, "They Might Be Giants").

Since the American chestnut was a valuable timber species, many dying and dead trees were harvested shortly before or after the arrival of blight in a given area. Lumbering created further disturbances in the forest, and logged-over areas were often ravaged by fires, soil erosion, and stream pollution. These, in turn, created open and barren conditions, and many sensitive, shade-loving herbaceous plant species also were lost.

As profound as these changes were for forest stands, the loss of the chestnut was even more devastating for wildlife. Unlike oaks, hickories and other nut-bearing trees, the American chestnut produced a highly reliable annual crop, providing high-energy food for birds, mammals and insects. Flocks of passenger pigeons, wild turkeys, blue jays, migrating bands of squirrels, mice, voles, raccoons, deer, elk, and black bear depended on the nuts. These animals, in turn, provided sustenance for the mountain lion, grey wolf, peregrine falcon, hawks, and other predators. Some of these species are now gone from eastern forests, in part as the result of the loss of the chestnut.

Many rural Appalachian families also suffered economic hardship as chestnuts began dying. For many mountain dwellers the wood, tannin extracted from bark, and sale of the nuts yielded vital cash. In a cruel twist of fate, chestnut trees began to disappear from many of these rural communities at about the same time that the Great Depression hit families the hardest.

Despite the devastation caused by blight, the American chestnut is not gone. In many areas old stumps still sprout (although less and less vigorously as time goes on) and these young stems grow rapidly when there is an opening in the forest. I have observed hundreds of chestnut sprouts per acre in West Virginia and in the mountains of western Maryland from old stumps when the dominant oak canopy was removed by clearcutting. These sprouts grow rapidly for several years until they, too, succumb to blight. Then the cycle is repeated.

Another mode of survival appears to be in the form of long-lived but suppressed seedling sprouts. Ecologist Frederick Paillet recently described several densely forested areas in Massachusetts and Connecticut where "inspection...indicated that the majority of living chestnut sprouts had originated as seedlings." His observations agree with those of Catherine Keever who, in the 1950s, also noted large numbers of seedling-origin trees in North Carolina. These "seedlings" are actually suppressed, nearly dormant, stems that may be 30 to 40-years old, yet rarely grow until canopy openings are created.

*Castanea dentata* continues to struggle against extinction. Presently it is restricted to an understory, shrub-like existence in densely forested areas or as rapidly emerging and dying clump sprouts in clearcuts. Its best hope

172

may lie with the dozens of scientists at major universities and research institutions who are working to breed resistant varieties of this species, or are attempting to vanquish the pathogen.

Unfortunately, this was neither the first nor the last example of such transcontinental pest or disease agents. Gypsy moth, Dutch elm disease, white pine blister rust, and the emerald ash borer are but just a few examples of the many other exotic assaults on vulnerable native plant populations caused by human meddling. Transcontinental movement of goods and people are likely to increase accidental introductions of pests and pathogens world-wide. If there is any good news, it is that, unlike a century ago, we are now aware of these possibilities and, perhaps, we can guard against them in the future.

# THEY MIGHT BE GIANTS

*A tree is a big plant with a stick up the middle.*
*-Colin Tudge*

Travelers heading southbound on Interstate 79 from Erie, Pennsylvania, are likely to observe an interesting springtime phenomenon near Grove City. Just north of where Interstate 80 bisects I-79, the climate and native vegetation are very different than just a few miles south of I-80. That's because the northern branch of the Jet Stream, that fast-moving air current that traverses the North American continent a few miles above the surface of the Earth, frequently seems to hitch a ride across I-80 in its rush toward the east. On the north side of the Jet Stream it is often cold and snowy, due to chilly air streaming down from Canada. But just a short distance south of the interstate, much milder air pushing up from the south creates a noticeably different environment. Thus, the Jet Stream acts as a de facto divider between north and south. This is reflected in the timing of the springtime emergence of grass, and the flowering of plants like tulips, crabapples, daffodils, lilacs and other early-blooming plants. Indeed, the very makeup of the plant life encountered in the landscape on the two sides of the Jet Stream seems very different.

One of the noticeable changes south of I-80 is the gradual appearance of one of the nicest tree species, *Liriodendron tulipifera*. I mention its scientific name right away because it is also known by several very confusing common names (see more about this problem in the essay "A Rose by Any Other Name...") Foresters refer to it as yellow poplar (it is not a poplar.) Other names include the lily tree (it is not in the Lily Family); the tulip tree (it is not related to tulips or to the tulip tree of Africa); or "whitewood," which is a meaningless term. Early settlers called it the "canoe tree," a term they acquired from native peoples. For the sake of simplicity, I will call this tree the "tulip poplar."

Members of the Magnolia Family, there are two species of *Liriodendron*: in addition to *tulipifera* in North America, the Asian species is *chinense*. Why would two very closely related species be found continents apart? This is not an unusual situation, especially among members of the Plant Kingdom. The most likely explanation can be found in the study of biogeography and in the theory of continental drift.

According to a textbook definition, biogeography is the "study of distribution of species and ecosystems in geographic space and through geologic time." Some of the early students of biogeography included the 18th century naturalist/philosopher Georges-Louis Leclerc (also known as Comte de Buffon, a favorite pen pal of Thomas Jefferson), Alexander Humboldt, and Alphonse DeCandolle, whose name is also mentioned in the story, "A Different Kind of Bank." The "father" of biogeography, however, is the 19th century naturalist, Alfred Wallace, who also nearly beat out

Charles Darwin as the most famous proponent of evolution. These and other scientists of their day noticed not only differences but also similarities between closely-related organisms separated not only by great distances but, in many cases, also by vast expanses of oceans. Most of these seemingly puzzling observations had to await the theory of continental drift.

The most often credited modern day proponent of the idea that continents move around ("drift"), was the German geophysicist, Alfred Wegener. When he published this concept in 1912 as "continental displacement," his hypothesis was soundly rejected because, at the time, there seemed to be no credible explanation for a process by which huge land masses could become "mobile."

It wasn't until the latter half of the 20th century that a mechanism for this movement of continents was proposed: plate tectonics. This concept includes the observation that the outer surface of the planet (the so-called "lithosphere") is not a solid, contiguous layer, but is made up of a series of large and small plates which are able to move with respect to one another. Sometimes they collide, sometimes they slide on top or below each other, and sometimes they drift apart. Data published in 2015 indicate that the plates are able to slide around so easily because their base is made of molten rock. Much like a film of water sitting on top of a layer of ice makes the surface more slippery, the liquid between the plate and its underlying mantle have a viscous layer to aid in the movement.

Paleontologists have proposed that at one time (about 200 million years ago) there was a single proto-continent, named by Wegener "Pangaea." Over time, this supercontinent split apart into increasingly smaller sections to become today's geographic location of continental land masses. Organisms that were in close proximity to one another on the large continent became separated when the breakup of the land masses occurred. Over time, closely related organisms acquired new genetic traits through mutations and other chromosomal changes and became unique species. There is considerable evidence from all scientific disciplines to support the theory of continental drift and plate tectonics.

In North America, the tulip poplar is found in the eastern half of the United States, ranging from Massachusetts to Illinois, and southward to northern Florida. It is mostly absent from the colder regions of the Adirondack Mountains and the Appalachian Plateau in New York state and in the northern tier of Pennsylvania, but can be found hugging the warmer shores of Lakes Erie and Ontario. And, occasionally, it can be found planted in yards and as city trees. In very cold regions, the tree is often damaged by late spring season frosts and snow, but in warmer climes and in fertile ground, it is one of the fastest growing tree species. In open areas with rich, well-drained soils its growth rate outperforms conifers and rivals the once-dominant American chestnut.

Many writers claim that the tulip poplar is the tallest tree in the eastern U.S., surpassing even the giant white pines of New England. It grows not only tall, but straight as an arrow.

In southern states, it can reach heights of nearly 200 feet (63 meters), with trunks 11 feet across. I have read that there is an estimated 350-400 year-old tulip poplar in New York City, near the Long Island Expressway. Dubbed the "Queens Giant," it was 134 feet (42 meters) tall in 2012. I have wanted to see this tree for quite a while now, and I hope that it is still standing at the time of writing this essay, so that I may see it one day.

By the time the traveler crosses from Pennsylvania into West Virginia on I-79, the tulip poplar has become a prominent member of the canopy of certain forest stands. Not tolerant of shade, any opening on good sites is quickly colonized by the tulip poplar (especially since one of its former competitors, the American chestnut, is no longer a major player in the eastern forest). Because of its rapid growth rate, it can quickly suppress competitors. Its beauty and importance in the forest ecosystem are reflected in the fact that the tulip poplar is the state tree of Indiana, Kentucky and Tennessee.

The wood is light-weight, light-colored, and doesn't warp or check if kiln dried properly. As mentioned earlier, Native Americans and early settlers used the wood in canoe building and for a variety of other medium-duty purposes. I have used it to make coat hanger pegs, railings, and a variety of other household items, because it's easy to work with and stains well in interesting grain patterns. Just be sure to pre-drill any nail or screw hole.

I first encountered the tree at the Cincinnati Nature Center in Milford, Ohio, some 48 years ago. Since I never saw this tree in Europe (although I have read that it is commonly planted in many European countries), I was intrigued by its beautiful tulip-like yellow flowers, its unusual leaf, and its prehistoric-looking, cone-shaped seed bearing structure. The leaf is shaped sort of like a red or black maple leaf, except that where the tip of the leaf should be, there is a broad, shallow concave indentation. Once seen, the leaf shape cannot be mistaken for any other species (see below).

The tulip poplar has become my favorite tree. At the risk of sounding morbid, my one request for my grave site is that someone plants a tulip poplar (or a white pine) seedling nearby.

# A PENNY FOR YOUR THOUGHTS

*A long-standing difference of opinion among zoologists
divides those who regard animals as totally predictable
machines, from those who regard animals as complex,
often mysterious organisms.*
*-Hubert and Mabel Frings*

The other day I was about to throw away an old, worn out work jacket but first I looked through the pockets. Amidst the lint, old tissues, dry pine needles, a couple of tarnished pennies and other flotsam, I found a carefully folded paper. I could barely make out the words penciled in years ago and long forgotten. After deciphering the faded lettering, I suddenly remembered the summer day when I wrote the words, "barn swallows," "groundhog," and "cat." They were to remind me of what I saw that afternoon.

I was weeding the garden on a peaceful summer day and stopped to watch the swallows flying in and out of the barn. I had a big barn. I had lots of swallows. Their cup-shaped mud nests were plastered all over the hand-hewn beams of the old structure. The birds were flying in and out of the hayloft through the open windows and the large sliding door. They had babies who were hungry. A groundhog was foraging in the yard, leisurely moving about in the field of grass and clover. The swallows ignored him as they busily went about the business of catching bugs mid-flight.

180

Then the cat came out of the house, angling over toward the barn. Even though he hadn't yet crossed the yard the swallows were instantly alarmed. First one, then a second, then a whole bunch more began an aerial assault of the cat. They dove at great speed toward the feline, like kamikaze pilots aiming for the deck of a ship. They pulled up a split second before colliding with their target. I watched in fascination as the cat stopped, turned, and began a panicked full-speed gallop away from the direction of the barn. The swallows succeeded in driving away the enemy!

How could this be? How did the swallows know to ignore the groundhog who, as a dedicated herbivore, represented no threat. How did they discern that the cat, a dedicated carnivore, may be a potential predator? Was some "decision making" process involved in the selection of those very different responses? Was this a learned behavior – as in young swallows watching the older ones' reaction – or was this an entirely innate, "instinctive" reaction? It was puzzling.

I also keep chickens. I purchased them as one-day-old chicks, reared them inside the house until they were old enough to go outside, and relocated them out into their coop when they had enough feathers and the weather got warmer. They had no prior experiences with older hens who could teach them anything. Yet, when Mr. Hawk begins to circle overhead, they abandon whatever they are doing and make a mad dash for their house. I have watched this countless times when I hear the shrieking of a hawk. The chickens' reaction is always the same: the hawk makes its appearance and they bolt for safety.

For the hens, this behavior must be instinctive, or inborn, or genetically programmed – whatever one wants to call it. They had no role models to learn from. And they also know not to fear geese, jays, crows, or even a turkey vulture flying over their yard. So, the chickens also make decisions? Is there a conscious thought process involved in their actions?

I have to confess that the study of animal behavior (also called "ethology") is not a topic that I know a lot about. I have several books on the subject and have read parts of them and numerous articles on ethology, but most of these simply describe the multitude of experiments performed without explaining how animals "know" how to respond. Other papers delve in great detail into the neurobiology of behavior, or relate evolutionary adaptive advantages to certain behaviors.

But no one really knows what an animal's decision-making process – if any – involves, or the level of cognition they possess. At best, one gets the idea from books that primates are "smarter" than rats, who are smarter than frogs, who are smarter than worms. Yet, I have seen earthworms in the process of copulation ignore me when I walk by them, but instantly decouple and disappear into their burrows when a hen gets too close. Are they reacting to size, shape, scent, or some other cue to decide who is harmless and who is about to eat them? Is their reaction based on the same instinct as the swallows' response to the cat, or is it something else?

Of course, I can't ask them these questions. Unlike Dr. Doolittle, I can't communicate with them.

The writer Henry Beston may have put it best in his classic 1928 book, *The Outermost House. A Year of Life on the Great Beach of Cape Cod*:

*"We need another and wiser and perhaps more mystical concept of animals. Remote from universal nature, and living by complicated artifice, man in civilization surveys the creature through the glass of his knowledge and sees thereby a feather magnified and the whole image in distortion. We patronize them for their incompleteness, for their tragic fate of having taken form so far below ourselves. And herein we err, and greatly err. For the animal shall not be measured by man. In a world older and more complete than ours they moved finished and complete, gifted with extensions of senses we have lost or never attained, living by voices we shall never hear."*

That thought is good enough for me!

# A DAMMED GOOD STORY

*Canada was built on dead beavers.*
*-Margaret Atwood*

We have new neighbors down the road. I haven't met them yet but I know that they are there. I think they are more like squatters than genuine land owners – there was no "for sale" sign on the property, no real estate agents, no notices of sale in the newspaper. I bet they are not going to pay taxes, either.

At first, there were just a few small clues that they moved in: a couple of small trees cut down, a small cluster of branches here and there, those sorts of things. The trees weren't cut, really; they were gnawed until they fell over. Then larger alder and pussy willow trunks and branches appeared in the big drainage pipe channeling the stream water under the dirt road. Now, they are cutting down bigger apple and aspen trees farther away and building a more formidable structure with wood, stones and mud. Eventually, I suspect, they will create a small lake and set about building a house and have a family. I think I know their kin – they live about a mile down the road and I have had encounters with them in the past.

I walk down to the stream several times a week to see what my neighbors are up to now. I guess, if I really wanted to meet them in person, I should go over there at night with a flashlight. But, somehow, I don't think that they would want to meet me face to face, even if I had a tin of cookies as a goodwill gesture. I am, of course, talking about the beavers that arrived in the neighborhood.

The North American beaver, *Castor canadensis*, is one of the most important animals in the human history of this continent. While most people would probably pick the horse or the buffalo for that label, the beaver is largely responsible for the rapid westward expansion of settlements by Europeans in the 17th and 18th centuries in what today are the United States and Canada.

The beaver had a central role in the territorial conflict between the British and French crowns, culminating in the French and Indian Wars of 1756-63. That's because the beaver has something that was highly coveted and extremely valuable back then. That something is its fur. In the 1600s through early 1800s, the principal use of beaver fur was for hat making. After the European beaver (*Castor fiber*) was hunted to near extinction, demand turned to its North American cousin.

Prior to trade with the Europeans, beavers were hunted by native North American tribes for meat, medicinal potions and warm clothing. Unlike the white settlers, Indians wore garments with the fur side of the pelt against their skin. Several skins were stitched together with sinew into skirts,

185

coats and other articles of clothing. Most native tribes harvested relatively few animals at a time, although there is a story of an eastern tribe massacring about 600 beavers in a macabre religious ceremony to drive away evil spirits. Even so, in pre-colonial times there may have been as many as 400 million beavers in North America.

With the arrival of European colonists, a brisk trade in furs ensued. Trading furs with Indians for trinkets did not satisfy the demand back in Europe or the avaricious appetite of the settlers for riches. More and more hunters and trappers began pushing into wilderness areas in pursuit of the beaver. During the height of demand, more than 200,000 pelts were shipped annually to European hat makers.

A well-made hat required several pounds of the soft "underhair;" the coarse guard hairs had to be removed and were usually discarded. The hat became a status symbol and the wearer of a high quality hat was afforded instant recognition of prominent social standing. Hats could be very expensive: at the peak of its popularity in the 17th century, an excellent hat could command the price equivalent of six months' wages for a skilled laborer.

Few individuals had higher hopes or a keener vision of the riches to be made from beaver furs than the two Frenchmen who founded the Hudson's Bay Company. Pierre-Esprit Raddison and his brother-in-law, Medard Chouart, Sieur de Groseilliers, approached the governor of New France (Canada) with a plan for setting up a fur trading partnership in Northern Canada. Their efforts were rebuffed by the governor and the two entrepreneurs eventually obtained financial backing from England, along with a Royal Charter.

In May, 1670, King Charles II of England granted possession of what amounts to one-third of today's Canada east of the Rockies to the new company, officially known as The Governor and Adventurers of England Trading Company in the Hudson Bay. The charter gave the company virtually unlimited powers to create and enforce laws, maintain an army and navy, and make treaties or wage war with native tribes. This enterprise eventually became known as the Hudson's Bay Company, and a much smaller version is still in operation today

The French, suddenly realizing that they let a lucrative opportunity slip away also coveted the land and its treasures. The British and their nemesis fought several wars over the territory and the fur trade, until a provision in the Treaty of Utrecht firmly placed control of the lands in the hands of Great Britain. Animosity and skirmishes continued, however, and the French were stoking anti-settler sentiment among the native peoples. The two countries and their Indian allies eventually went to war again in 1756.

Although the Hudson's Bay Company traded in all kinds of furs, including lynx, snowshoe hare, muskrat and others, it was the beaver that made it rich and powerful. The firm's influence continued to grow after it acquired its main rival, the North West Company in 1820. The Hudson's Bay Company then had monopoly over all of Canada, excluding

Quebec. Its western possessions were eventually lost when in the 1840s, the United States claimed legal rights to lands south of the 49<sup>th</sup> parallel.

The decline in the popularity of the fur hat did not ease trapping pressure on the beaver. Other garments and accessories made from beaver fur continued to be popular and the pelts still commanded high prices. The animals also were hunted for food and the purported medicinal value of castoreum, an oily substance that the beaver uses to waterproof its coat (castoreum is not the same as castor oil, which is obtained from the seeds of the castor bean plant).

Of equal or greater importance in the decline of beaver populations, rivers and streams were altered and channeled for flood control or water-powered mills. Wetlands were drained for agricultural use, and forests were logged bare, displacing the animal from its natural habitats (please see the story, "Rebirth," elsewhere in this book).

By the early 20th century, *Castor canadensis* had become nearly extinct in many parts of North America. For example, the last known beaver in Union County, Pennsylvania, was killed in 1912. Legislation passed in many states and Canadian provinces eventually led to the protection of the animal from indiscriminate and wanton destruction. Efforts began to re-establish beaver populations by shipping them to states from which they had disappeared.

The beaver has made a remarkable recovery in the past 100 years. Protection from unlicensed trapping, the abandonment of millions of acres of farmland, and the renewed appreciation for the value of wetlands have allowed this large rodent to reassert itself into many areas. Ecologists often refer to the beaver as a "keystone" species. The term indicates that, in suitable habitats, the activities of the animal will determine the ultimate fate of the other inhabitants of the ecosystem.

Beavers moving into a new territory quickly build dams on small streams and brooks. Some experiments suggest that the sound of running water is a powerful motivating force for the construction of the earth-mud-stone- and tree-limb structures. Water is also channeled through a series of canals, which serve to float building materials for the construction of the dams. The rising water levels behind the dam inundate flat areas and lead to the formation of ponds or lakes. If the low-lying area is initially forested, the trees die after a few years because their roots suffocate from the lack of oxygen. This gradual tree decline displaces some residents but creates habitats for others.

Some years after the construction of a pond, aquatic plants appear along with a new fauna of invertebrates, amphibians, and fishes. There are dramatic shifts in animal species outside the water, too. Ducks, geese, herons and other water fowl replace terrestrial birds. Muskrat, mink and occasional otters take advantage of the watery environment.

Beaver dams can become enormous in size. The largest known dam in Alberta, Canada, is almost 3,000 feet long and several hundred feet high. Beavers actively repair dams to keep water levels at the desired height. Water serves as a protection from predators and it is also a means of locomotion and transport of food.

A few years ago I owned a farm down the road from my present residence. There was a beautiful old 3-story barn near the stream and I loved this old structure (it's the same barn I wrote about in the story titled "A Penny for Your Thoughts"). When the grand-uncles and grandparents of the present-day beavers moved in they set about damming up the creek. I watched as water levels rose perilously close to the foundation of the barn. Armed with pick, shovel and saw, I sought to dismantle the dam. It was hot, sweaty and dirty work and I got a million tiny cuts from the marsh grass in the water, and another million or so bites from insects attracted to my heat and sweat. I'd open a big cut in the dam; by the next morning it was repaired. We played this game day after day.

Standing knee-deep in the murky water, I couldn't help but muse over my armchair ecologist days in Philadelphia decades earlier when, looking at pictures of beavers in *Natural History* or in *National Geographic*, I thought how "cute" they were. Those warm and fuzzy feelings quickly disappeared in the middle of a dammed up stream.

Now, I think back on these experiences as I watch the present generation hard at work. The criss-crossed tree branches and trunks are getting higher and more elaborate. The water level keeps rising. They haven't started on their lodge, yet, but it's only a matter of time. I suspect that the town will eventually send a crew to try to dismantle the tangle of branches and open up the flow of water through the pipe again. I wish them luck!

Post Script: The town did take action. Shortly after I wrote this story, the beavers were removed by a licensed trapper and heavy equipment was brought in to dismantle the as yet unfinished dam. But there is a big lodge upstream with a growing family. Parents give marching orders to move out when the youngsters are about 2 years old. Sooner or later a new generation of beavers will return to the culvert.

Post-post script: In 2016, a new generation of beavers moved into the culvert. These were also trapped and relocated and their partially constructed dams were torn down. The creek is flowing again, but for how long…

# A HAWKISH TALE

*...general widespread control of raptor*
*populations is undesirable.*
*-J.J. and F.C. Craighead*

For the past 20 years I have been driving the 150 mile round-trip distance between Allegany County, and Rochester, New York, at least once a week. Most of this distance is driven on Interstate 390, an easy-to-drive, straight road that traverses a glacier-scoured, gently rolling landscape. The highway is usually sparsely traveled, leaving plenty of time to enjoy the scenery without having to worry about large trucks or crazy drivers.

Twenty years is a long time. The years have allowed for observing changes in land use, buildings, and farming practices. I mention this because I have seen a huge increase in the red-tailed hawk population during these two decades. Nowadays, it is not unusual for me to see 15-20 of these birds while I'm driving. And these, of course, are only the ones that I can spot along the road. Back in 1996, I would have been lucky to spy one or two hawks along the same drive.

My observations are not statistically quantifiable. After all, I am counting birds perched on tree tops or in flight, while I'm driving at 65 miles per hour. But I can tell that they

192

are mostly red-tailed hawks from their size, the color of their plumage, and the shape of their wings in mid-flight. It is more difficult to identify them when they set up their sentry positions in a tree but, even if not all of the hawks are red-tails, unquestionably there are more hawks these days.

The red-tailed hawk (*Buteo jamaiencis*) is a "generalist" feeder. In other words, it eats many things, including reptiles, birds, and small mammals. Its staple food, however, consists of field mice. Hawks also eat carrion and road kills are, unfortunately, all too common.

Most ornithologists agree that red-tailed hawk populations are apparently increasing in many parts of their range. Experts propose that this is not due to reduced mortality from pesticides; unlike other raptors, these hawks were apparently not decimated by the wide-spread use of DDT and other toxic chemicals after World War II (please see the essay about the Colorado potato beetle elsewhere in this book.) Rather, the increase is most likely due to changing land-use patterns and the adaptable nature of these birds.

Red-tailed hawks prefer a landscape with small patches of woods, rather than large tracts of forested land, for hunting and nesting. They like isolated, tall trees that provide lookout perches, and open agricultural fields that provide plenty of prey (mice, voles, rabbits, etc.). Because of the available food supply, coupled with far less average snow cover in the past decades, and supplemented by plenty of road kills (raccoons, foxes, possums, etc.), some of the formerly migratory hawks have been increasingly reluctant to fly south in all but the worst winters.

On January 15, 2017, I counted ten red-tailed hawks between Dansville, Livingston County and Rochester, a distance of 54 miles. And a week later, there were 11 of them (possibly some of the same birds that I saw before) over that same stretch of I-390.

An additional contributing factor to the hawk's increase in numbers may be reduced hunting pressure. Sometimes called the "chicken hawk," because it will prey on domestic fowl, the red-tail can occasionally decimate flocks of chickens. Since the number of farms keeping chickens is not what it used to be, shooting of the hawks has decreased.

I have personal experience with hawks taking my chickens. A number of years ago I acquired a small flock of Bantam hens. Several of these small birds were quickly taken by red-tailed hawks. I did put netting over the chicken yard to discourage the plunder, but I would never have thought of shooting one of these hawks.

Like raccoons, turkey vultures, deer, and the red fox, the red-tailed hawk is a good example of a highly adaptable organism that has exploited its opportunity to live close to human populations. Thus, human activity not only eliminates some species, but also alters ecosystems in many other ways, including inadvertently encouraging the increase of some species.

If we accept the premise that the environment selects who lives and profits, and who declines and perishes, then we can look at ourselves as an environmental force, just like climate change, plate tectonics, or some other evolutionary force. Humans have shaped the world's landscape ever since we left our home in Africa. This change has accelerated in the last few centuries as populations of people have skyrocketed and as technological advances have allowed for greater and greater levels of ecosystem disturbance. It is tempting to think about what the world will look like in a century or a millennium, but I will leave that speculation for the science fiction writer.

# DEAR DEER

*The deer brought to bay, or plunging into the flood
from the pursuit of the Indian, armed with bow and
arrow, seemed wanting to realize the savage landscape
as it appeared to the first settlers of the country.
-Thomas Nuttall, 1854*

I look out the window at the old, gnarly apple trees that
were part of an orchard a long time ago. There are still a few
shriveled yellow apples on branches that are too high for the
deer to reach. They have eaten all the ones they could get,
plus my azaleas, Forsythia and everything else. I see half a
dozen animals right now, pawing at the snow to get at the
grass. They are hungry and there are a lot of them in this
neck of the woods. The entire yard is littered with their
droppings. I have had to put all the bird feeders on pulleys
that propel the containers 10 feet in the air; otherwise, the
deer head butt the feeders within their reach to scatter the
seeds.

It is estimated that in the U.S. there are about 30 million
deer. No one knows for sure; even individual states'
estimates are not very accurate. According statistics I found
on The Outdoor Channel Digital Networks, Texas (4
million), Mississippi (1.7 million), Michigan (1.4 million),
New York and other states (each with about 1 million)

represent the largest estimated deer populations. In Mississippi, there are more than 650 deer for every 1,000 people, and West Virginia is not far behind at 500 per 1,000 humans. Some states have seen a near doubling of their deer populations within the last 20 years.

These animals don't just cause losses in agriculture and landscaped property. There are nearly 1.2 million deer-vehicle collisions in the U.S. annually. These accidents result in 200 human fatalities, 29,000 injuries, and cost 1.66 billion dollars in property damage, emergency services, medical bills, and road cleanup each year.

Why are we experiencing such huge increases in white tail deer populations? There is no single answer to this question but the life history (autecology) of the deer itself, current forestry practices, changing land use policies, and changes in population control methods (both "natural" and "artificial") that kept deer herds in check in the past all contribute to the increase.

White-tailed deer (*Odocoileus virginianus*) are large, plant-eating mammals that require about 3-8 lbs (1.5-4 kg) of food per day. Well-fed deer are prolific. Generally, two fawns are produced yearly by each female, and reproduction levels may boost the population by 20-30 percent annually. Some of these fawns will not, of course, survive through their first year, especially in harsh winters when food supplies are low or unobtainable. Many inexperienced yearlings (as well as adults) die from injuries sustained in automobile collisions.

Deer browse herbaceous vegetation, such as grasses, clover, annual flowering plants, and weeds in the summer, but are able to switch their metabolism to live on buds and young twigs of trees and woody shrubs in the winter. They are not especially choosy about their winter browse; over 100 species of woody plants have been recorded consumed by deer. They also damage orchard trees, nursery stock and even Christmas trees. Whenever they can, they will eat crops left in the field or stored in corn bins. Deer typically do not wander too far from home and continue to enlarge their populations for as long as food is available.

In the old-growth forests first encountered by European settlers, tall canopy trees dominated and shaded out woody shrubs and young seedlings, and the forest floor was open and free of browse material. Fragmentation of these once-contiguous forests created many openings where ground pine, young tree seedlings, shrubs, and sub-canopy tree species, like dogwood, could proliferate. These, then, are quickly devoured by deer.

Silvicultural practices, such as clearcutting, create immediately open areas that are quickly populated by thousands of seedlings. Logging companies often clearcut large tracts of forests, or use a modified method known as seed tree cutting. These companies may then proudly point to the fact that they replant cut over areas with up to a million seedlings. However, as far as the deer is concerned, this simply represents a huge dinner table. Keeping deer out of these areas is impractical, since about the only method of preventing feeding is to enclose the area with an 8-ft-tall fence, preferably electrified. Not even the richest forestry companies can afford to erect such fencing over thousands of hectares of land. Thus nourished, deer populations expand exponentially.

Contrary to popular belief, most of the forested areas in the northeastern part of the U.S. are not owned by large companies but are small parcels (10-200 acres; 4-80 hectares) belonging to individuals. These small woodlots are even more fragmented and are interrupted by open fields, farms, housing developments and roads, and towns and cities where deer roam freely. Deer love these varied habitats and the food they provide. Thus, winter starvation, once a major natural population control method, has been eliminated from many areas. There may be trouble ahead for the deer, however: forests in the eastern U.S. are once again aging and are returning to a mature, closed-canopy state. These second-growth forests will, then, provide less food for the large numbers of deer.

In addition to the lack of winter forage, a number of native predators once kept the white-tail in check. Wolves, panthers and lynx took many old, sick or very young deer. With the disappearance of the carnivores, deer are not culled from the population by these predators. Native Americans and early settlers hunted deer but took relatively few animals for food and clothing sewn from their skin.

Eventually, the eastern states were populated by ever-increasing numbers of newcomers who used any and all means of hunting. Deer were lured to feeding stations and salt licks to be slaughtered; hunting dogs and buckshot were used; later, spotlights were employed to track this dawn and dusk/nocturnal species. According to published reports by the Pennsylvania Game Commission, the numbers of deer were so low in the Commonwealth in the early 1900s, that about 1,200 animals were "imported" from other states between 1906 and 1924. Gradually, states outlawed the indiscriminate hunting methods, and strict quotas and seasonal dates were established. Once protected, deer populations quickly rebounded.

Many states derive handsome revenues from controlled hunting licenses and the taxation of products used by hunters. For example, the impact of deer hunting in Texas is estimated to be about 2 billion dollars a year. In New York, hunting contributes about $ 1.4 billion to the state's economy. I recently saw a poster on the internet, sponsored by the South Carolina Department of Natural Resources. It said: "Hunters: Help Control Coyotes and Save Our Deer!" I doubt very much whether the South Carolina DNR cares greatly about the white-tail. I am certain that their officials are thinking dollars and cents when they encourage people to kill coyotes who could at least make a dent in deer populations. But, the senseless hunt of the coyote is a different topic for a different time (or read Barbara Kingsolver's *Prodigal Summer*).

At one time one of the major predators of white-tail deer were cougars. A research article published by University of Alaska and University of Washington scientists in the

200

August, 2016, issue of *Conservation Letters* estimates that if cougars were reintroduced into the eastern states, deer populations could be reduced by 22 percent. This decline would prevent 680 injuries, 5 deaths, and $50 million property damage annually, because there would be fewer deer-vehicle collisions. The scientists estimate that a single adult cougar could kill about 260 deer over its adult lifespan. Yes, cougars do take livestock, too, but estimates put this loss at only about $2.3 million per year. However, major opposition by hunting clubs, animal rights advocates, livestock producers, government agencies, and other groups, would likely derail any proposal to bring back the cougar to the eastern part of the U.S.

In the absence of major predators, other than humans, deer have become what ecologists call the "keystone" species in many areas (I mentioned the beaver, too, as a keystone species in some areas). This means that they have the largest impact on the makeup and ultimate fate of the ecosystem because they control what plants will become the dominant vegetation type in the future. For example, in many forested areas in the northeast, native tree species are in the process of being replaced by a monoculture of young beech trees because the deer don't like to browse those seedlings. In my own small woodlot, there are hundreds of beech seedlings and saplings. Yet, despite the many mature sugar maples that produce prolific seed crops every year, there is not a single maple seedling to be found. They are devoured as soon as deer find them.

I have mixed feelings about deer. They are handsome and graceful animals. At the same time, they are very destructive and I must be vigilant at all times to keep them away from our plants. My best weapon is my fearless little pug dog, Sophie. She weighs only about 12 pounds, but she bravely charges deer ten times her size. The deer scamper, jump and run, but they are back quickly. I gave up trying to grow some plants, like azaleas or tulips. It takes an 8-foot-tall metal fence to keep them out of my garden.

Unhappy as they make me, I also must realize that they were here first, and I need to learn to share "my" land with them. And, unless a major population control factor (like chronic wasting disease or some other infectious agent, or a predator like the cougar or wolf) comes along, their numbers are not going to get smaller any time in the near future.

# WHAT IS THE LIFE OF A MOUSE WORTH?

*And God blessed men, and God said unto them, Be fruitful*
*and multiply, and replenish the earth, and subdue it;*
*and have dominion over the fish of the sea, and*
*over the fowl of the air, and over every living*
*thing that moveth upon the earth.*
*-Genesis 1:28*

As the weather turns colder in late September and early October there begins an annual migration of mice, voles, shrews, and squirrels into my basement and outbuildings. They are looking for shelter and warmth. But let's face it – for me they are a nuisance, or worse.

Part of my house was built in the 1840s, and the old stone walls of the cellar have many tiny cracks through which small rodents can enter. I have poured much concrete to seal these crevices but they always seem to find more. Mice and voles can and do chew into pink wall insulation, cardboard boxes, and assorted stored items, but my biggest worry is that they will gnaw the yards and yards of electric wires in the cellar and cause a fire.

In the barn and storage sheds mice, gray and red squirrels set up household and, in addition to chewing, their pungent urine ruins tents, tarps, camping equipment, boards, shelving, and many other items. Their sharp incisors can even slice through heavy duty plastic storage boxes in which they then set up their nests.

To combat them, I use traps. No, not the capture-and-release kind. I use several sizes of snap traps baited with peanut butter. These traps are effective. But every time I remove the lifeless, crushed body of a mouse or shrew or squirrel I feel a pang of guilt and remorse. I look at one of these mammals and realize that, although they don't look like me, their bodies are basically the same as mine. And I'm certain that they felt pain and may have suffered for some time before succumbing to the injuries that the snap of a metal bar had inflicted.

OK, so mice and squirrels are mammals and I – also being a mammal – may have some affinity with them and even affection (?) for them. But if I think about other animals, like frogs or beetles or chickens or worms, they are also very much like me at the cellular and molecular level. Their nerve cells and mine use the same sodium and potassium ion gates to propel electrical impulses and, thus, they must feel sensations like pain the same as I do. Their muscle cells use the same sliding protein filaments as mine to do their work; they have the same cellular structures to make energy; and their DNA is made up of the same chemicals as mine. In other words, all animals are related and that includes us, human primates, as well.

Some biologists make a distinction between animals that are "sentient" creatures, and life forms that supposedly do not have conscious awareness ("insentient"). This distinction is based, theoretically, on the presence or absence of a nervous system. Ok, but where exactly to draw the line? The earthworms I write about in "Red Wigglers" do have nervous systems, although these may be anatomically different from the nervous system of a mouse. So, then, do earthworms have "conscious awareness" or the "power of perception" associated with sentient beings? I think so. Certainly, an insect, like a mosquito, has a nervous system and it is very good at finding me by using its abilities to sense my heat or gradients of carbon dioxide output, or some combination of factors. Does it make it into a sentient being? And, if so, do I have the right to maim it or kill it?

I am no Jainist or someone like Gandhi. I confess elsewhere in this book that I loathe slugs. I also don't like and fear snakes (it is called ophidophobia) and I'm convinced that my early religious upbringing has a lot to do with that. I don't kill snakes but I do kill the slugs that destroy my garden. I am also not averse to "getting rid of" flies, mosquitoes and other pesky insects. And, obviously, I snap trap some mammals.

I often wonder what, if any, moral right I have to make this distinction between different animals. I am especially confused because I absolutely love our three dogs and I grieved for weeks when one of my beloved cats died a few weeks ago. So, I began to read some books on this subject. What I have discovered so far is that I am not alone in this quandary and, in fact, most people who have thought about this subject feel pretty much the same confusion as I do.

In my early teens my friends and I would go to this abandoned quarry where rain water pooled into small ponds, which were full of frogs. Our goal was to find suitable rocks to hit and kill as many of the frogs as we could. In my readings I have found that in most surveys 20 to 30 percent of people (both men and women) have, at one time in their lives, perpetrated cruelty on various types of animals.

In the 1970s I worked in animal facilities at several universities in Ohio and Pennsylvania. Part of the job was to "dispose" of mice, rats, rabbits, and gerbils that were no longer "needed" for experiments. There were days when we "euthanized" (Killed!) hundreds of these creatures. I eventually quit the profession of laboratory animal technology because of this.

By the time I moved to West Virginia in the early '80s, I was appalled and disgusted when I would see posters advertising the annual "rattlesnake roundup," where hundreds of these reptiles were massacred. I consider this a senseless and barbaric act, especially since rattlesnakes will leave you alone if you leave them alone, and they are handsome and useful animals. So, there has been a gradual change in my thinking about the callous ways with which we humans treat other animals.

Maybe, I am a hypochrite of sorts, too, because I eat meat. I don't ever kill my chickens, but I eat store-bought poultry and eggs regularly. I also eat red meat occasionally and, although I love and respect pigs, I like "pork" products. I have rationalized this by saying that, oh, well, we are omnivores, after all, and by reminding myself that animals that eat other animals certainly don't think about this or feel remorse. At least, I don't think that they do.

Hal Herzog, in his book *Some We Love, Some We Hate, Some We Eat,* calls human relationships with other species "messy." My thinking on the role of humans in the environment we share with other organisms has been "evolving" (aaah, see my discussion on the use of this word in the essay "Little Monkeys on the Grass..."). From wanton killing of caterpillars and frogs as a child, to mourning (yes, that is the correct word for me) the killing of mice who invade my basement, has been a 50 plus year journey. Maybe, I'm just getting sentimental in my old age. Or, maybe, I have begun to appreciate life in a way that I would have been unable to do so as a teenager.

So, what is the life of a mouse worth? I don't have the answer to my question. But I think it is worth asking the question just the same to keep in perspective that life is amazing, and that all organisms have an intrinsic value even if we can't always attach monetary or moral or practical value to them.

# WINTER MUSINGS

## Concord, Massachusetts, 1992

*I prefer winter and fall, when you feel the*
*bone structure of the landscape*
*-Andrew Wyeth*

I always tell anyone who cares to listen that autumn is my favorite time of the year. Today, however, I opt for winter. It is a peaceful, slate-colored day with gently falling snow. Already, most of the trees and the ground are covered with a virgin layer of white; the boulders scattered throughout the woods have yet to acquire their coats. My old work boots make no sound as I leave diamond-shape patterns of the thread of their soles behind. It is incredibly quiet now with the newly-fallen snow acting as insulation against the clatter of civilization.

I come to a fork in the trail and, choosing a steep, winding path, I begin my ascent. Nothing moves until out of the corner of my eye I spy a squirrel. It is a handsome and well-fed creature; its silver-tipped fur sparkles even on this cloudy afternoon. We play a game. I step to one side – he goes halfway around the tree and stops head down. I step the other way and he moves in the opposite direction now head pointing up toward the sky. We slow-dance like this for a minute or two until he tires of it. Up the maple trunk he goes and with a graceless jump he dangles from the branch of a neighboring hemlock tree. The branch swings wildly and then he is gone.

I resume the climb. I come to the top of the hill, huffing and puffing, and look down at a pond below. A kettle hole, created thousands of years ago by a big chunk of glacial ice that melted away slowly in the huge depression it made. Some piece of ice this must have been – the pond is a good hundred feet below the top of the ridge. The last time I was here, during springtime, there was a deafening croak of spring peepers in that water, with the walls of the kettle hole creating a natural amphitheater to amplify the sound. Now, all is peacefully quiet. The frogs are still there but they sleep their deep slumber until the springtime ritual of mating begins again. Frozen cattail stubble sticks out of the ice at the edge of the pond. The water in the middle is perfectly still, taking on the gray color of the sky and blending into green near the shore where hemlocks see their own reflections.

The snow comes straight down now in huge, wet flakes in the windless afternoon. I enjoy the scent of the woods in the air but then, almost immediately, I contradict this feeling by lighting a cigarette. The sulphury odor of a poorly lit match leaves an acrid stench in my nostrils. Even as I enjoy the nicotine coursing through my body, I am furious with myself: *Why can't I give up these damned things?* I try to sit on a rock but its jagged edge is uncomfortable. I clear the snow from a small depression in the ground and stretch out a piece of plastic to sit on; it's an old bread wrapper that I brought along for this purpose. I smoke quietly, lost in thought. Here comes another squirrel. This one is a rufous color and I can tell he is curious about me. He stops at what he must consider a safe distance and we eye each other for a while. *I wonder what he is thinking. Does he wonder about what I'm thinking?* A screeching blue jay frightens us out of our reverie. The squirrel runs and I stub out my cigarette in the snow. The wet butt and the plastic wrapper go in my back pocket – I have my own woods etiquette. I debate whether to climb down to the pond but choose, instead, another trail. This one is level and circles around the ridge top. There are oaks here with their dry brown leaves still clinging to branches. I walk up to a copse of gnarly white pines whose trunks are stained with mostly dry sap. A lot of insect injury on the stems; borer holes that the pines are trying to plug with resin. I scoop up a small amount of still-fresh pitch on a fingertip and enjoy its piney scent.

Suddenly, I come to the edge of the woods. Beyond here it's mostly farmland. The ploughed rows are neatly arranged in alternating black and white lines where the snow has settled on the tops of the furrows. My mind wanders back to another place and time, to another glacier-scoured landscape just south of Columbus, Ohio. There, glacial moraine was deposited to form small knolls in an otherwise almost completely flat terrain. My friends and I walked the farmer's plowed field on a day very much like this. We were supposed to be hunting but none of us had much desire to kill anything. If I remember correctly, we didn't even load the guns.

This time, my memories are interrupted by a bird of another kind, a small prop plane landing in a nearby air strip.

The spell is now broken. Civilization has intruded, even though I knew all along that I was only about a quarter mile from the nearest highway. I am just about a mile away from Walden Pond, as the crow flies, where Henry David Thoreau spent a year or so contemplating and writing about nature. But, of course, that was nearly 150 years ago, and I'm sure his reverie was not interrupted by the clangs, bangs and whirrs of modernity that I'm experiencing right now.

I quickly glance at my watch. I still have about two hours before I have to pick up the kids from school but I no longer feel peaceful here. It's time to go. I turn around, light another smoke, curse myself again, and head back to the parking lot. There will be other times to cherish these woods if these cigarettes don't kill me first.

********

Post Script: I quit smoking in 2008, after a nearly 42-year-old addiction to nicotine. Although I had tried everything from acupuncture to hypnosis to medication, nothing seemed to work until one day I said, "I had enough." That was the end of that addiction.

# GIORGIO

*In the Big Rock Candy Mountain,*
*There's a land that's fair and bright,*
*Where the handouts grow on bushes*
*And you sleep out every night.*
*-Harry McClintock*

My wife said, as she handed over the plate of breakfast bagels and eggs, "I wonder how Giorgio can manage in this snow." Looking out the dining room picture window we saw the thick blanket of heavy snow that fell overnight. The graceful branches of the Norway spruce tree bowed under the weight; the lifeless branches of the mock orange bush leaned drunkenly to one side. A dozen or so chickadees crowded the feeder while snowbirds on the ground were busy picking up seeds that the birds above were sending their way. I, too, had thought of Giorgio earlier that morning. This snow came on top of another 10 inches already on the ground and it made for difficult walking. Especially for someone like Giorgio, the opossum.

He showed up one day a few weeks earlier when I spied him on the porch eating dry food that I put out for our cats. He scampered away as soon as I opened the door but then he was back in a few minutes. I watched from inside the house, letting him have his fill of the tasty morsels. After that he came just about every day and we left various treats for him: leftover meatloaf, a bit of chicken, apple cores, stale bread, things like that. He wasn't a fussy eater and consumed everything we offered. He would come, eat, and disappear.

He came back often enough that I named him (after a Giorgio Armani advertisement hanging on our refrigerator door) and we thought we had a resident in our woods. The cats did not bother him when he came on the porch; they would turn their heads and pretended not to see him. One day, after a heavy snowfall we saw him coming from the woods and it was a mighty struggle for him to reach the house. He kept sinking in with each step, and he tried short hopping maneuvers, using his tail as both rudder and to keep from sinking into the soft layer of snow. We watched him from the window, wishing that we could scoop him up and help him, all the while knowing that we couldn't.

I have lived in many states with cold climates and I have always marveled how the opossum can tolerate our harsh winters. They are seemingly so poorly-equipped to be able to do so. The opossum – sometimes just called possum – is known to biologists as *Didelphis virginiana*. It is the only known representative of the marsupials, or "pouched" mammals, in North America. Most of the rest of the members of the Family Didelphidae live in tropical or sub-tropical areas of Central and South America, although there is considerable fossil evidence that these oldest of mammals originated on the North American continent. The name opossum is said to be a corruption of the Algonquin Indian word "apasum."

The size of a cat but with much shorter legs, the possum's coarse, shaggy fur is thin, its ears and rat-like tail are bare, its feet and soft, fleshy toes are ill-equipped for digging an underground burrow, and it does not hibernate. In other words, possums should, by all indications, freeze to death in our winters. That's obviously not the case, though individuals with frostbitten tails or ears are frequently encountered.

So, how do animals like possum, deer, foxes and turkeys avoid the brutal cold and icy winds of northern latitudes? There is no single answer to this question but many kinds of aquatic and terrestrial animals have sophisticated systems for keeping warm. Of foremost importance is to keep the vital organs of the torso and head functioning. Down, hair, feathers, fat and thick skin are employed for this purpose. Core body temperature is maintained at or near a "set-point" (a narrow optimum range) to keep the heart, lungs, kidneys, and other organs fully operational.

Since there is a potential for a lot of heat loss, animals in cold climates tend to have larger body sizes than those of the same species in warmer areas (I wrote about this so-called Bergmann's Rule in the essay "A Horse of a Different Color") because bigger bodies have smaller surface-volume ratios (think of the analogy of a large piece of wood versus kindling, where the smaller pieces are used to start a fire due to their total exposed surface area). Shivering is another way to generate some body heat.

Extremities are, of course, part of the body but these are often able to be kept at lower functional temperatures. Known to physiologists as "spatial heterothermy," animals from whales to seagulls to land mammals conserve heat energy this way. Even in humans, whose internal organs and head are kept at the set-point temperature of close to $37^0$ Celsius ($98.6^0$ Fahrenheit), the arms and legs may become much colder at $31-34^0$ C ($88-93^0$ F).

Humans cannot maintain low temperatures in hands and feet for too long. We must wear protective clothing or find a place whose temperatures allow for warming these body parts. Most warm-blooded animals that are active throughout the cold months of the year don't have the option of wearing gloves and coats – certainly wild animals don't. Therefore, there must be some other means of protecting limbs, wings, tails, ears, noses or any other protruding body parts of soft tissues, while at the same time conserving precious energy.

The ingenious system they have is a marvelous example of nature's engineering know-how in the form of a heat exchange system, called countercurrent exchange. The blood vessels in exposed body parts are arranged so that arteries carrying warm blood from the torso are in close proximity to veins returning cold blood back into the body. In this way, most of the heat of the arterial blood is passed to the veins instead of dissipating from the surface of the limb. Many animals even have a shunt system, so that on days when the animal does need to cool, the countercurrent system can be bypassed.

Environmental engineers charged with designing energy-efficient buildings recently adopted a similar concept of high-capacity heat exchange systems. Whereas the human-made temperature regulating systems are operated by computer-controlled thermostats, a similar function in animals is performed by the hypothalamus of the brain (please see the story, "A Long Winter's Sleep.")

What may be surprising is that not only warm-blooded animals (birds and mammals) can regulate their body temperatures. We may think of insects as helpless, torpid creatures in cold weather – and to be sure, most of them are. But some moths, wasps and bees generate enough heat to be able to fly at $5^0$ C ($41^0$ F).

No, I did not think about all of this over breakfast on that cold, snowy winter morning when we were lamenting Giorgio's absence. But the wheels started turning in my head thinking about our little friend and all of the other creatures out there in that deep snow.

We didn't see Giorgio that day or the next. In fact, Giorgio never came back again. We kept assuring ourselves that it is not uncommon for possums to hang out in one locale for a short period of time and then move on. They are solitary, transient hobo-like creatures who keep traveling, perhaps looking for that "Big Rock Candy Mountain" that the old hobo song talks about.

I hope Giorgio found it.

# A ROSE BY ANY OTHER NAME...

*In the fields of observation, chance favors a prepared mind.*
*-Louis Pasteur*

I wrote about a lot of different organisms in this book. Each time I described some creature, I tried to mention its scientific name to avoid possible confusion as to which organism I was referring. You might ask: "Why would there be possible misunderstanding if I simply used the common name of a plant, bacterium, fungus or animal?"

The common names of organisms are often misleading. Pineapple is neither pine nor apple, prairie dogs are ground squirrels instead of canines, and horse chestnut is not even remotely related to true chestnuts, other than the fact that they are both woody plants. Some creatures are known in various places by vernaculars. For example, the common groundhog (*Marmota monax*) is also known as the woodchuck, marmot, red monk, or whistle pig. The latter name is also given to the Pacific Northwest cousin of the groundhog, *Marmota caligata*, and to the totally unrelated South American Guinea pig (*Cavia porcellus*).

The common lawn weed, broadleaved plantain (*Plantago major*), has several dozens of different common

English names. It is also known by a dozen local names in French-speaking countries around the world. To muddy the picture even further, several unrelated plants are also called plantains.

With hundreds of different languages, and thousands of different local names in existence, biologists all over the world have to be able to communicate with one another and must know whether they are discussing the same organism or talking about two different ones. The Greek philosopher and naturalist, Theophrastus, appears to have been the first individual around 300 B.C., to make an attempt of making sense of this confusion.

The "father" of modern classification [taxonomy], however, is Carolus Linnaeus (born Carl von Linne), the 18th century Swedish botanist who, in his 1753 book, *Species Plantarum,* began to use a system of naming organisms that is still in use today. Known as the binomial (two names) system of nomenclature, Linnaeus brought order to chaos by naming each living organism identified up to that time with a genus and species designation. The genus, a larger grouping of closely related organisms, is then followed by the species (the word "species" is both singular and plural). The species designation is specific for a given plant or animal that is distinct from other kinds within the same genus. In this way, the North American flowering dogwood, *Cornus florida*, is related to, but is not the same plant as the Asiatic dogwood, *Cornus kousa.*

Linnaeus knew nothing about genes or chromosomes, because discovery of these took another 200 years. Rather, he described different organisms based on their physical characteristics and living habits. Based on Plato's and Aristotle's concepts of the "fixed" forms of plants and animals, each species was believed to have been created in a "perfect" form by a Creator. All organisms belonging to that genus and species were considered immutable, and all variations of form were merely imperfections of the grand design. It took 19th century scientific and philosophical thought to challenge these long-held assumptions. All branches of biology contributed to our understanding of the similarities and differences between different types of life forms, and classification systems are still "evolving" today.

There wasn't always universal harmony among specialists in taxonomy. While many accepted the general idea of the binomial system, some naturalists sharply disagreed on the actual mechanisms of naming organisms. For example, the International Botanical Congress, meeting in Vienna in 1905, adopted a set of international standards for naming plants. This was quickly rejected by the American delegation, whose members set up their own system. A nasty and bitter 30-year-long debate ensued between the two factions, and there is at least one anecdote about two well-respected scholars and gentlemen who clobbered each other with their walking canes after a heated disagreement. Despite the infighting and squabbles, the Linnaean system is still used for classifying all living organisms, including bacteria. The only entities for which this system did not seem to work are the viruses, although taxonomy is beginning to be employed for this group, as well.

Today, biologists often describe a species not by its outward appearance, as Linnaeus did, but by subtle but significant differences in their genes that prevent interbreeding between two species (here I should say, however, that even the species concept has begun to be challenged by some taxonomists). Nor is a scientific name cast in stone. Once adopted, the Latin binomial is subject to change as new taxonomic information becomes available. The names of some fungi, for example, have been changed as many as four or five times since they were originally described and named. This is not only confusing, but it has caused many students studying taxonomy to consider ripping out their hair strand by strand.

Since Linnaeus used mostly Latin (or Latinized Greek) names, a little knowledge of these languages goes a long way toward deciphering the seemingly bizarre names given to organisms. Many of these tongue-twisters are actually highly descriptive of some characteristic possessed by an organism.

Colors are especially popular in the species names of plants; most art students will quickly recognize words such as *caerulius* (sky blue), *cinnabarinus* (cinnabar red), and *cupreous* (copper-colored). Those well-versed in medical terminology will no doubt know the meaning of *sanguineous* (blood red) and *vitellinus* (dull reddish yellow).

Some names are somewhat grandiose as well as descriptive. The species name, *atropurpurea,* denotes "terrible" (i.e. very intense) purple; *macrophylla* means large leaf; *exaltata* stands for very tall; and the genus name *Phytophthora* means "horrible destroyer of plants." The marine invertebrate, *Nautilus,* derives its name from the Greek word for "sailor", and the head of the water flea, *Cyclops,* resembles the mythical one-eyed giant.

Some names are not very exciting or even descriptive or useful. Species names such as *virginianus, canadensis,* or *carolinensis* were probably adopted because the species were first found in those localities or, perhaps, because the person doing the naming wished to honor their home state or country. The white-tailed deer, *Odocoileus virginianus,* is not only native to the Commonwealth of Virginia, but it is found in just about every state except the arid southwestern United States.

Other names were coined to salute eminent naturalists: *Browallia* (John Browall), *Thunbergia* (Carl Peter Thunberg), *Brownia* (Patrick Brown), and *Listeria* (Joseph Lister) are a few examples.

Then there is the oriental barberry, *Berberis julianae,* which was named by a botanist in honor of his wife, Juliana.

And the January 20, 2017, edition of *The Wall Street Journal* reported that the scientists who discovered a new species of moth named it *Neopalpa donaldtrumpi* because of the 45th President's "distinct hairdo, resembling the yellow scales atop the moth's head."

What may be actually surprising is how many of these tongue-twisters have become everyday names of organisms. Take, for example, *Amaryllis, Coleus, Rhododendron* and *Coreopsis.* While it is unlikely that most of us will routinely refer to dandelions as *Taraxacum officinale,* scientific names are useful and, with a little detective work, can even be fun.

# A FISH STORY - NOT!

*You have brains in your head*
*You have feet in your shoes*
*You can steer yourself any*
*Direction you choose.*
*-Dr. Seuss*

In this book of essays I wrote about birds, mammals, amphibians, bacteria, viruses, various types of invertebrates, plants and fungi. I did not write much about fish. The first and most important advice given to would-be writers is to write what you know about. I don't know a lot about different kinds of fishes.

To a greater or lesser extent we are all prisoners of our history. In my case, personal history, family history and the history of my birthplace, Hungary, all have conspired to keep me from caring about fish. Both of my grandfathers were soldiers in the Austrian-Hungarian army during World War I. Both were captured and both spent several years as prisoners of war in Russia. My father, who was not a soldier but was nevertheless taken by the Soviet Red Army as prisoner in the waning days of World War II, also spent four years in Siberia. All of them survived on stale bread, fish and occasional cabbage broth as their staple diet.

I never knew my paternal grandfather because he died of a heart attack before I was born. My father, terrified of once again being captured by the Russians during the Hungarian Revolution of 1956, fled for Austria. My maternal grandfather was the dominant male in my early life. After all those years of eating fish in prison camp, he swore he would never look at it, smell it, or eat it again. As a result, fish was "verboten" in our house. My grandmother, who was quite piscivorous, would sneak an occasional can of sardines into the house and I watched, admittedly with great disgust, as she devoured the little silver fishes with the cold bulging eyes. To this day, I don't like the smell of fish, not even tuna.

It's not even that I'm not interested in these animals. In my former college office, I kept a sealed glass tube with the skeleton of a dog shark, and this always made for an interesting conversation piece. The shark's skeleton is made entirely of cartilage, and it is a true ocean-dweller. In my classes (I still teach on a part-time basis) we discuss that the shark's body fluids are in balance with the salt content of ocean water, whereas bony fishes living in the ocean have the same basic problem with drinking salty water that freshwater creatures or we land animals have.

So, it's not the lack of curiosity that keeps me form exploring fishes. It's a deeper, more visceral aversion to their smell, their looks and textures, imprinted on me from early experience. It makes me wonder whether I will ever be able to overcome my dislike for fish or whether I should even try. But as Scarlet O'Hara says in *Gone with the Wind,* "After all...tomorrow is another day."

It took me several years to put this volume together. Perhaps, by the time I write another set of stories about the wonderful and fascinating world of biology, I can report that I had salmon steaks for dinner the other night.

\*\*\*\*\*\*\*\*\*\*\*\*\*\*\*\*

I had a lot of fun researching and writing the various stories. Life is amazing and beautiful.

\*\*\*\*\*\*\*\*\*\*\*\*\*\*\*\*

Although I don't fool myself into thinking that this book will rival *Harry Potter* in popularity, I thank you for taking the time to read these essays.

# ADDITIONAL READING

### INTRODUCTION
Mole, B. 2013. Material inspired by dragonfly wings bursts bacteria. Science News. Dec. 28. pg. 11.
Sanders, L. 2011. Helping artificial limbs to feel real. Science News. Feb. 26, pg. 10.

### KARTOFFELKÄFER
www.PotatoBeetle.org

### REBIRTH
Douglas, Marjory Stoneman. 1988. The Everglades:River of Grass. Revised Ed., Sarasota, FL: Pineapple Press
Keddy, P. 2010. Wetland Ecology: Principles and Conservation. 2nd. Ed. New York: Cambridge Univ. Press

### WHEN HEARING IS SEEING
McFarland, D. 1993. Animal Behaviour. 2nd Ed. Essex, U.K.: Longman Scientific
Barbour, R.W. and Davis, W.H. 1974. Mammals of Kentucky. Lexington, KY: The Univ. Press of Kentucky
Caswell, B. and C. Doeller. 2013. 3D mapping in the brain. Science, 19 April, 279-280.

### SHOULD WE STAY OR SHOULD WE GO
Elman, R. 1974. Hunters Field Guide to the Game Birds and Animals of North America. New York: A.A. Knopf Publ. Co.

## *QUO VADIS,* BIRD FLU?
Falkow, S. 2012. The lessons from Asilomar and the H5N1 "affair." mBio 3(5):e00354-12.
Taubenberger, J. and Morens, D. 2006. 1918 influenza: the mother of all pandemics. Emerging Infectious Diseases 12(1). 8 pgs.

## THE CELL FROM HELL
Wood, P. April 12, 2008. Pfiesteria hysteria? CapitalGazette.com

## "EVERY FACE GATHERED PALENESS..."
Johnson, S. 2006. The Ghost Map. New York: Riverhead Books

## A DIFFERENT KIND OF BANK
Pollan, M. 2001. The Botany of Desire. A Plant's-Eye View of the World. New York: Random House
Siebert, C. 2011. Food Ark. National Geographic, July 108-131.
Vavilov, N.I. 1992. Origin and Geography of Cultivated Plants. Translated by Doris Love. Cambridge, U.K.: Cambridge Univ. Press. (from the original Russian text written in the 1930s).

## A LONG WINTER'S SLEEP
Hanney, P. 1975. Rodents: Their Lives and Habits. Vancouver, Canada: Douglas David &Charles LTD
Saey, T.H. 2012. Lessons from the torpid. Science News. Feb. 25, pgs. 26-29.

## A STONE, A WINDOWPANE AND THE NIGHTLY NEWS
Bower, B. 2010. Effects of botox go beyond the face. Science News. July 31, pg. 8
Lewis, C. 2002. Botox cosmetic: A look at looking good. FDA Consumer 36:11-13

## A HIDDEN WORLD IN PLAIN VIEW
Price, P. 1975. Insect Ecology. New York: John Wiley &Sons
Snodgrass, R. 1967. Insects. Their Ways and Means of Living. New York: Dover Publ., Inc.

## THE PRIVATE LIVES OF FERNS
Dott, R.H. and R.L. Batten. 1981. Evolution of the Earth. New York: McGraw-Hill Book Company.
Gould, J. L. and W.T. Keeton. 1996. Biological Science. New York: W.W. Norton & Co.

## FLY, FLY AWAY
Alcock, J. 1993. Animal Behavior. 5th ed. Sunderland, MA: Sinauer Assoc., Inc.
Long, M. 1991. The secrets of animal navigation. National Geographic 179:70-99.

## SPRING SYMPHONY
Pough, F.H., C.M. Janis and J.B. Heiser. 2002.Vertebrate Life. 6th Ed. Upper Saddle River: Prentice Hall

## AUTUMN DELIGHTS
Arora, D. 1986. Mushrooms Demystified. Berkeley, CA: Ten Speed Press

**I'M DRAWN TO YOU**
Blakemore, R. and Frankel, R. 1981. Magnetic navigation in bacteria. Scientific American 245:58-65.
Merrill, R. 2010. Our Magnetic Earth: The Science of Geomagnetism. Chicago, IL: The University of Chicago Press
Witze, A. 2013. Spinning the core. Science News. May 18, pgs. 26-29.

**"LITTLE MONKEYS ON THE GRASS..."**
Moore, D.D. 2005. Most Americans tenetative about origin of life explanations. The Gallup Poll.
www.pollingreport.com/science/htm.
Newport, F. 2009. On Darwin's birthday, only 4 in 10 believe in evolution. The Gallup Poll.
www.gallup.com/poll/114544/Darwin-Birthday-Believe-Evolution.aspx

**A HORSE OF A DIFFERENT COLOR**
McFadden, B. 1999. Fossil Horses: Systematics, Paleobiology and Evolution of the Family Equidae. Cambridge, U.K.: Cambridge Univ. Press
Saey, T.H. 2013. Ancient horse's DNA fills a picture of equine evolution. Science News. July 27, pgs. 5-6.
Secord, R. 2012. Evolution of the earliest horses driven by climate changes in the Paleocene-Eocene Thermal Maximum. Science 335:959-962.

**"ONE IF BY LAND..."**
Haddock, S.H.D., Moline, M. and Case, J. 2010. Bioluminsecence in the sea. Annual Review of Marine Science 2:443-493.
Mihail, J. and Bruhn, J. 2007. Dynamics of bioluminescence by *Armillaria gallica, A. mellea* and *A. tabescens.* Mycologia 99:341-350.
Tucker, A. 2013. Light fantastic. Smithsonian 43 (11): 50-59.

## A STICKY SITUATION

Ehrenberg, R. 2009. Worm-inspired superglue. Science News. Sept. 12, pg. 8.

Simon, H. 1976. Snails of Land and Sea. New York: The Vanguard Press, Inc.

Watnick, P. and Kolter, R. 2000. Biofilm, city of microbes. J. of Bacteriology 182:2675-2679.

## RED WIGGLERS

Appelhof, M. 2006. Worms Eat My Garbage. How To Set Up and Maintain a Worm Composting System. 2nd. Ed. Kalamazoo, MI: Flower Press Publ. Co.

Steiner, R. 2011. Agriculture Course: The Birth of the Biodynamic Method. Forest Row, U.K.: Rudolf Steiner Press

## BIGGER THAN A BREADBOX

Smith, M., Bruhn, J. and Anderson, P. 1992. The fungus *Armillaria bulbosa* is among the largest and oldest living organisms. Nature 356:428-431.

Volk, T. 2002. The humongous fungus ten years later. Inoculum 53:4-8.

## NATURE'S ALCHEMISTS

Pessarakli, M. (Ed.). 2016. Photosynthesis. 3rd. ed. Boca Raton, FL: CRC Press

Tobin, A.J. and J. Duscheck. 2001. Asking About Life. Pacific Grove, CA: Brooks/Cole-Thomson Learning

## "ONE OF THE MOST USEFUL AND BEAUTIFUL TREES..."

Freinkel, S. 2007. American Chestnut. The Life, Death, and Rebirth of a Perfect Tree. Berkeley, CA: Univ. California Press

## THEY MIGHT BE GIANTS
Hugh Johnson's Encyclopedia of Trees. 1990. New York:
Portland House.

## A PENNY FOR YOUR THOUGHTS
Milius, S. 2012. Face smarts. Science News. Oct. 6, pgs. 20-
23.
Stokes, D. and Stokes, L. 1983. Stokes Guide To Bird
Behavior. Vol. I-III. Boston, MA: Little, Brown and Co.

## A DAMMED GOOD STORY
Merritt, I. 1987. Guide to the Mammals of Pennsylvania.
Pittsburgh, PA: The University of Pittsburgh Press
Muller-Schwarze, D. 2011. The Beaver: Its Life and Impact.
Ithaca, NY: Comstock Publ. Assoc.
Perkins, S. 2009. Ancient beavers did not eat trees. Science
News. Nov. 21, pg. 10.

## A HAWKISH TALE
Craighead, John J. and Frank C. Craighead, Jr. 1969. Hawks,
Owls and Wildlife. New York: Dover Publ., Inc.

## OH, DEER
Rue, III, L. 2004. The Deer of North America. Guilford, CT:
Lyons Press
Thomas, E. 2009. The Hidden Life of Deer. New York:
HarperCollins

## WHAT'S THE LIFE OF A MOUSE WORTH?
Dunayer, Joan. 2001. Animal Equality. Language and
Liberation. Derwood, MD:Ryce Publishing
Herzog, H. 2010. Some We Love, Some We Hate, Some We
Eat. New York, NY: HarperCollins Publishers.

## GIORGIO
www.opossum.org

**A ROSE BY ANY OTHER NAME...**
Bailey, L. 1963. How Plants Get Their Names. Mineola, NY: Dover Publ.
United States Department of Agriculture - Natural Resource Conservation Service (USDA - NRCS) Plants Database. plants.usda.gov

# ACKNOWLEDGMENTS

I would like to thank Brian Quinn, Bob Curry and Sherry Volk for helpful suggestions upon reading rough drafts of the early versions of this manuscript. Sherry also scoured the *Giorgio the Possum* edition of the text line-by-line and corrected my spelling, punctuation, and other errors. My wife, Debra, and my youngest son Steven went through this volume with a fine tooth comb. As there are still likely to be mistakes, blame my poor typing skills and repeated text changes after earlier proof-reading efforts by Sherry, Deb and Steven.

I thank Ms. Annette Burdett with advice on formatting and the intricacies of word processing, and her cheerful and encouraging early morning attitude. In fact, I have had the good fortune of working with four remarkable women at Alfred State College: in addition to Annette, Ms. Mary Scholla, Ms. Amy Chan, and Ms. Roxana Sammons have always been there for me over the years.

I gave copies of some of the stories to my students in biology classes. Polite as these students were (or, perhaps because they erroneously assumed that their comments would affect their grades), their reviews were generally favorable and their suggestions extremely useful.

I thank Toni, Matt, Felicia, Bob, Steven, Deb and Alex for their love, support and encouragement.

I cherish the memory of my mom who always supported me in all my doings, even when she didn't agree with my decisions.

I have been privileged to work with living things for most of my career and I hope that there are other planets out there somewhere that also have the right environment to support life – in whatever form that life may exist.

.

www.ingramcontent.com/pod-product-compliance
Lightning Source LLC
Chambersburg PA
CBHW021404170526
45164CB00002B/497

*9 7 8 1 5 4 2 5 2 6 2 6 5 *